CODE NAME
GINGER

HARVARD BUSINESS SCHOOL PRESS

Boston, Massachusetts

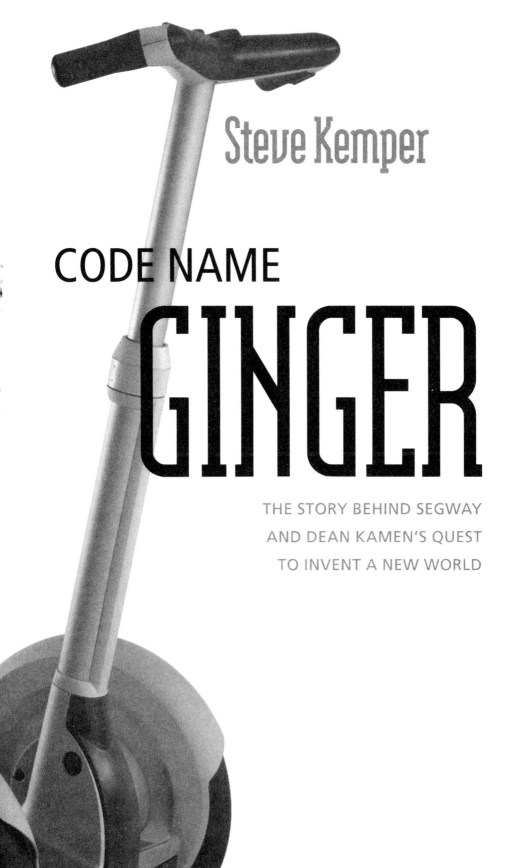

Steve Kemper

CODE NAME
GINGER

THE STORY BEHIND SEGWAY
AND DEAN KAMEN'S QUEST
TO INVENT A NEW WORLD

07 06 05 04 03   5 4 3 2 1

Library of Congress Cataloging-in-Publication Data
Kemper, Steve.
  Code name Ginger : the story behind Segway and Dean Kamen's quest
to invent a new world / Steve Kemper.
     p. cm.
  ISBN 1-57851-673-0
  1. Scooters. I. Title.
  TL410.K46 2003
  629.22'7—dc21

                                        2002155932

The paper used in this publication meets the requirements of the American National Standard for Permanence of Paper for Publications and Documents in Libraries and Archives Z39.48-1992.

For Jude, my gyro

CONTENTS

Cast of Characters   ix

PROLOGUE	First Impressions	1
ONE	In Development	9
TWO	A Dream with a Deadline	31
THREE	CEO Mode	45
FOUR	Frogs	57
FIVE	Winter Solstice	77
SIX	Lions and Angels	89
SEVEN	Tuck and Roll	107
EIGHT	Deaned	119
NINE	The Slip	149
TEN	No More Shakespeare	165
ELEVEN	Sanctioning Bodies	183
TWELVE	The Path of Minimum Pain	199
THIRTEEN	All Hat, No Rabbits	217
FOURTEEN	Everything Is Connected All the Time	235
FIFTEEN	West Coast Ambush	251
SIXTEEN	Fallout	263
SEVENTEEN	The Leak	283
EIGHTEEN	The Reveal	297
EPILOGUE	Castles in the Air	309

Author's Note   315
Acknowledgments   317
About the Author   319

DEKA

Dean Kamen — *founder and owner*
Bob Tuttle — *Dean's chief financial adviser and deal maker*
Mike Ambrogi — *Dean's longtime right-hand man*

The iBOT Project (Fred)

Benge Ambrogi and Kurt Heinzmann — *engineers*
Lucas Merrow — *general manager*

The Ginger Project

Tim Adams — *CEO and president*

ENGINEERING

Doug Field — *director*
Bill Arling — *lead electromechanical engineer*
Jim Dattolo — *software engineer*
J. D. Heinzmann — *motor drive development*
Phil LeMay — *lead electronics engineer*
Mike Martin — *lead engineer for structural integrity*
John Morrell — *lead controls engineer*
Ron Reich — *lead mechanical engineer*
Kevin Webber — *design engineer*

INDUSTRIAL DESIGN

Scott Waters — *lead designer*
Tao Chang

MARKETING

Mike Ferry – *director*
Tobe Cohen
Matt Samson

MANUFACTURING AND OPERATIONS

Don Manvel – *director*
Chip MacDonald – *plant-manager-in-waiting*

REGULATORY AFFAIRS

Brian Toohey

FINANCE

Scott Frock

INVESTORS AND WOULD-BE INVESTORS

Paul Allaire – *CEO and chairman of Xerox, retired*
Jeff Bezos – *founder and CEO, Amazon.com*
John Doerr – *partner, Kleiner Perkins Caufield & Byers*
Jack Hennessy – *director, Private Equity Division,*
    Credit Suisse First Boston
Steve Jobs – *cofounder and CEO, Apple and Pixar*
Vernon Loucks – *CEO of Baxter Healthcare, retired*
William Sahlman – *professor, Harvard Business School*
Michael Schmertzler – *U.S. cohead, Private Equity Division,*
    Credit Suisse First Boston

CODE NAME
GINGER

First Impressions

ill it change the world, as some have predicted? Charles Gibson posed that question to millions of people on the morning of December 3, 2001. For a week, *Good Morning America* had been promising to reveal the secret invention known to the world by its code name, Ginger, or by the mysterious acronym, IT.

Ginger had been "the topic of a global guessing game," said Gibson, the show's cohost. Reporters and pundits, incited by hyperbolic praise of Ginger from Apple's Steve Jobs, Amazon's Jeff Bezos, and superstar venture capitalist John Doerr, had written feverish stories in a dozen languages after word of the secret project had leaked out eleven months earlier. Internet forums devoted to the mystery had buzzed with hundreds of thousands of messages. Enthusiasts had debated every possibility and investigated every clue. On the search engine Lycos, Ginger broke into the top 10 most popular subjects, neck-and-neck with Britney Spears.

Internet sleuths discovered the patents for Ginger almost immediately, but the public hungered for a technological miracle and preferred fantasy. People transformed Ginger into a device for time travel, teleportation, or magnetic levitation. They imagined a hovercraft, a jet-pack, and an inertial thruster (based on the acronym IT). A writer in Thailand foresaw "a flying super tuk-tuk." After the terror of September 11, 2001, a rumor started that Ginger had been covertly deployed in Afghanistan to defeat Osama bin Laden. All this loony speculation had convinced some commentators that Ginger was a hoax, and had inspired comics to spoof it in the media and on the Web.

Good Morning America's camera shifted to a spot-lit object covered by a white sheet. An unimpressive T-shaped silhouette showed through the shroud. A plinking piano tried to sound portentous. The scene evoked Geraldo Rivera more than Thomas Edison.

Gibson's cohost, Diane Sawyer, swerved between sarcasm and hype. The silhouette reminded her of "either a giant asparagus or a vacuum machine." But she also referred to Ginger's creator, Dean Kamen, as "the legendary inventor."

The legend himself stood next to her, looking like he needed an espresso. He hated to get up early. He also looked uncomfortable, or maybe just embarrassed by the stagy buildup, which, with characteristic bluntness, he later termed "cheesy."

At five-foot-six, Dean was several inches shorter than Sawyer. His dark wavy hair rose in a semipompadour above a narrow face. He was slender (the tag on his Levis said waist 30, inseam 28), and looked younger than his fifty years. He wore jeans, work boots, and a denim shirt. That might seem informal for the debut of a machine that could change the world, but it was Dean's uniform, and he wore it whether meeting with his head machinist, the CEO of a Fortune 100 company, or the president of the United States. (When the weather turned cold he added an army surplus jacket whose pockets clanked with small tools that set off alarms in airports and at the White House.)

Dean had done a design freeze on his wardrobe sometime in the early 1970s as a teenager on Long Island. Filling his closet with jeans and blue work shirts was efficient and practical. He liked to think of himself as a down-and-dirty engineer who wouldn't hesitate to grab a greasy wrench or a soldering iron. He liked working with tools and knew how to run all the complicated machines in the shop at DEKA, his engineering research and development firm in New Hampshire.

That was the kind of engineer he hired, too—someone who liked to make things and get grubby whenever necessary. A visitor couldn't tell the difference between DEKA's engineers and machinists. Dean liked that. But the truth was that he didn't often get his hands dirty at DEKA anymore. He was too busy putting out fires and trying to raise money.

Dean didn't give any more thought to popular culture than he did to fashion. Movies? A waste of two hours. Novels? Why settle for make-believe when you could be pondering Isaac Newton's *Principia* or machining a spur gear? Television? You must be kidding. The newest music? No thanks, he'd stick with the oldies. Sporting events? Don't get him started on that one. He claimed that he had never bought a newspaper in his life. "My hobby is thinking," he often said.

Consequently, he had escaped the celebrity lint that clutters most people's minds. He once met Andy Warhol but didn't know who he was. He went to a fancy dinner and sat between two people he'd never heard of, Warren Beatty and Shirley MacLaine. At a White House conference on health care he sat next to a woman who talked a lot but made no sense. When she left for a moment, Dean turned to someone and said, "*She's* an expert on health care?" "Well, no, that's Barbra Streisand." A former professional basketball player once inquired about renting Dean's island off the Connecticut coast. Dean had never heard of him and kept referring to him as Jabbo. It was Kareem Abdul-Jabbar. "Dean, you're a failure at popular culture," Jeff Bezos, the founder of Amazon.com, once said to him. "You fill your head with useless things like the laws of thermodynamics."

Over the past year, during which the media had catapulted him from respectable obscurity into the national spotlight, he had been called everything from a genius to a publicity hound. He had become a celebrity engineer, a phrase that sounded self-contradictory, or at least unlikely. To other people, that is. Dean believed that engineers and scientists deserved to be cultural heroes, as famous as athletes and movie stars. He had sometimes courted fame himself, but he hadn't expected the abrupt and alarming way it swept him up in January 2001.

Dean had made millions by inventing new technologies and leasing them to bigger companies for a royalty, but Ginger excited him so much that he had decided to keep it for himself. He started a new company to handle the invention's development, marketing, and manufacture. Though he knew almost nothing about these things, he brimmed with self-confidence. To raise money for his vision, he had seduced some of the most prominent leaders of corporate America, defied the conventions of venture capitalism, and mustered the gall to turn down money from Steve Jobs and Jeff Bezos—yet he ended up with $90 million in funding and nearly 80 percent of the new company. "I did it the way I usually do," he had told the Ginger team. "I figure out what I think is fair, and then make sure everyone compromises and does it my way."

A year and a half later, here he was next to Diane Sawyer on *Good Morning America,* watching a sheet rise from Ginger. Standing naked in the spotlight, the machine looked anticlimactic: two wheels, a platform, and a T-bar.

After a pause, Gibson asked what it did. "It's the world's first self-balancing human transporter," said Dean, his Long Island accent still evident after nearly twenty years in New Hampshire. Gibson asked why it didn't topple over, as physics seemed to demand. "That's the *invention,*" said Dean. "It does what a human does. It has gyros and sensors that act like your inner ear. It has a computer that does what your brain does for you. It's got motors that do what your muscles do for you. It's got tires that do what your feet do for you."

Sawyer had been silent, staring at Ginger. "I'm tempted to say, 'That's *it?*'" she blurted. "But that *can't* be *it.*"

She didn't understand what she was looking at. She couldn't see the 90 percent of Ginger's story hidden below the surface, and didn't know about the intense, ragged process that had moved this invention from idea to marketplace. She couldn't see the flashes of discovery, the setbacks, the impassioned engineers, the high-flying investors, the multimillion-dollar misjudgments, the intrigues that had ousted some of the project's original leaders, the triumphs large and small.

That 90 percent is the subject of this book. I was with the engineers as the machine took shape, and I shadowed Dean as he scrambled to put together the financing. I stayed in Dean's house, discussed the project with him over late-night dinners, flew with him in his jet as he stitched together plans for a new international industry. I watched him outmaneuver and outsmart everyone, and sometimes sabotage himself or his team. I was there when he inspired his engineers and when he infuriated them. But by the time the world saw Ginger on *Good Morning America,* I had been exiled. That's a small part of the story, too.

In early 1999, Dean called and asked me to come up to DEKA in Manchester, New Hampshire. He wanted to show me a secret project, the biggest thing he had ever done, and to make me a proposal. Who doesn't love a mystery? Besides, Dean's pull can be as irresistible as the moon's. I went to New Hampshire.

We had known each other since 1994, when I wrote a story about him for *Smithsonian* magazine. His mix of inventive genius and brazen salesmanship had fascinated me. Enormously talented, with ambition to match, he had already invented many revolutionary medical devices, including the first drug infusion pump and the first portable dialysis machine. He held dozens of patents in areas ranging from helicopter design to computerized heating and cooling systems. He seemed destined to burst into wider public view.

When we first met, he had recently started FIRST, a robotic competition that teamed corporate engineers with high school students to inspire kids to consider careers in science and engineering. That year I watched forty-three teams compete in a New Hampshire high school gym. It was fun and small. "In a couple of years we'll be holding the national championship at Epcot Center in Disney World," he said.

I smiled at this bravado, but a few years later FIRST was ensconced at Epcot, where it grew into the center's biggest event. By 2002, more than 20,000 kids on 600 teams were competing in 17 regionals leading up to the championship, and Dean had wangled expensive sponsorships from dozens of Fortune 500 companies. (In 2003, 800 teams are competing in 23 regionals.) His penchant for outrageous overstatement was matched by accomplishment.

So I went to New Hampshire in February 1999 to listen to Dean's pitch. We sat in his corner office in one of the old mill buildings he had renovated. Like most great salesmen, Dean is a gifted raconteur, and that day he built suspense like a master. He began to narrate—no, *perform*—the events and insights that had led to the secret projects he called "Fred" and "Ginger."

He took me to see Fred first, using a key-card to enter a secured testing facility called Easy Street. Fred (today called the iBOT) was a super-duper wheelchair that could roll through gravel and sand, go up curbs, and even climb stairs. Most amazingly, it could rise up on its back wheels and roll along in balance. That's what always made the disabled test drivers cry, the simple dignity they reclaimed by "standing up" and looking people in the eye. Dean called Fred the world's most sophisticated robot. Like most things at DEKA, it had originated in his brain. Fred was extraordinary, and Dean's belief that it would change the lives of thousands of disabled people seemed plausible. He was developing the wheelchair for Johnson & Johnson, the project's funder.

But what he *really* wanted to show me was Ginger, which used similar technology for a very different end. In his analogy, Ginger would be General Motors; Fred was merely the sideline of ambulances and police cars that GM occasionally made.

Before taking me to see Ginger, he bombarded me with facts and statistics about cars and pollution and energy consumption, about clogged cities and traffic jams and the burgeoning population of the Third World. All of these, he said, would be dramatically affected by Ginger. He described the invention as "a personal transportation system" that didn't use fuel, took up no more space than a pedestrian, and weighed just 50 pounds. Like Fred, it could go up hills, down stairs, and through mud, gravel, and sand. Even if Ginger flopped among the general public, which he couldn't imagine, he would still sell millions of them—for starters, to the country's 2 million delivery people, not to mention all the golfers and infantrymen, whose numbers he also cited.

More important, the machine would transform cities and habits of transportation. It was clean and cheap. It ran on a few cents of battery-powered electricity per day and wouldn't be expensive. And oh, by the way, it was fun.

"It's going to change the world," he said, dead serious.

We walked to an adjacent mill building and climbed to the fourth floor. A sign on the locked door said Ginger Project. Dean key-carded us in. We passed a conference room into a lab whose tables were crowded with tools, gauges, and oscilloscopes, then continued into a long, narrow room. Blinds covered the windows to thwart prying eyes. A small fleet of brightly-colored electric bikes and scooters stood against one wall. The wave of the future, said Dean sardonically, and pointed to one that was Lee Iacocca's great new thing. Ginger, said Dean, would make them all obsolete. He walked to where a small machine leaned against another wall.

Ginger. Like Diane Sawyer nearly three years later, I couldn't believe it. *That's* the world-shaker? After Dean's buildup it made a drab first impression. Two toy wheels flanked a thin metal platform with a metal T-bar rising from it. Compared to the sleek candy-colored scooters, Ginger looked as plain and stern as a schoolmarm among fashion models.

Dean flipped a couple of switches, stepped onto the platform, and leaned forward. Ginger darted off. When he leaned backwards, the machine reversed. He pirouetted Ginger in place like a spinning figure skater; the turning radius was zero. Ginger zipped around the room like a hummingbird, and as quietly. Once animated, this schoolmarm was beautiful.

I itched to ride it. Dean instructed me to trust the machine to balance and showed me the thumb switch for turning. I stepped on and tentatively leaned forward. Ginger scooted ahead. The machine felt alive. Within moments I was comfortable enough to go faster by leaning harder. Ginger had seemed mysterious when Dean was on it, but riding it myself felt natural as well as exhilarating, like gliding over a sheen of surf on a skim board. The machine seemed to read my mind and respond like a powerful muscle. I couldn't stop grinning and didn't want to get off.

Dean hustled alongside, talking without pause. "It's your personal magic carpet," he said. He expected Ginger to become a common verb—let's ginger to work, ginger to the store, ginger in the woods. The machine would be to cars what the personal computer had been to the mainframe. Ginger danced so lightly that after a rainstorm it didn't leave tread marks on freshly raked dirt. A military guy who had seen the machine said it wouldn't even trigger land mines. Older people and those with crippling infirmities would love Ginger for restoring their mobility. A couple of high-powered outsiders had told Dean that the impact of Ginger on the twenty-first century would be as big as that of the car and the TV on the twentieth. "Who's going to want to walk?" he said with typical hyperbole.

He talked about how one technology inspires others and about how people, even engineers and technologists, can be blind to new applications for familiar technologies. Western Union, whose telegraph wires crossed the country, had turned down the chance to buy Alexander Graham Bell's patent for the telephone. "What use could this company make of an electrical toy?" Western Union's president had asked, thereby becoming infamous for making one of the nineteenth century's worst business blunders.

People might dismiss Ginger the same way, but Dean didn't think so. "I may be wrong," he said, "but I know I'm not." He was paranoid—his word—that his secret would leak out. He worried that a behemoth such as Sony or Honda might get wind of it, put two hundred engineers on it, and shove him aside. "I hope you don't talk in your sleep," he said.

He had been concentrating on Fred for several years, but Ginger's day was coming. For the first time, he wasn't going to turn his invention over to a bigger client. He planned to form and run a new company, separate from DEKA, that would design, refine, finance, and manufacture Ginger, with himself as principal owner. To compensate for his inexperience, he had hired a senior executive away from Chrysler to become Ginger's CEO and president. He had also appointed a lead engineer and told him to start putting together a development team. Ginger was about to bloom.

Dean considered the invention so important that he thought its progress should be documented in a book. That's why he had invited me up. There was one catch: Since the project was secret, I would have to work on spec and hope to interest a publisher somewhere down the road. In return, I would have complete access to Dean and the project. He sold me.

In August 1999 I began spending at least two days a week with him and the Ginger team. As I watched Dean deal with engineers, marketers, investors, and suppliers, my appreciation of him grew. So did the uncomfortable thought that he might not like the book taking shape in my notes. He was accustomed to control and unambiguous admiration, and expected a statue with no bird shit on it, despite the pigeons everywhere.

Neither of us anticipated how the story would unfold. Even in the months after Dean kicked me out of the project, it was apparent in our phone conversations that a part of him never stopped wanting this chronicle about the invention that he believes will change the world. But in the end, a bigger part of him wanted control.

This is Ginger's story, though not as he would tell it.

chapter one

In Development

1951–1995

Rockville Center, Long Island, a small, pleasant suburb thirty-eight minutes by train from New York City, isn't known as a cradle of invention. Dean was born there in 1951 and endured its educational system through high school. Like many restless, energetic boys, he was an indifferent student. School bored him. Once, in elementary school, his teacher told the class that every number divided by itself was one: 7/7 = 1, 9/9 = 1, and so on. Dean liked the clean certainty of that. But what about zero divided by zero? "One" didn't seem right, so he raised his hand. The question irritated his teacher, who accused him of not paying attention. That was generally true, but not this time, so Dean retorted that her math made no sense. She called Dean's mother, herself a teacher, who told Dean to apologize. He refused. He knew he was right.

In seventh grade, Dean's teacher complained to his parents that he must be cheating in math, because he got the right answers but didn't show any calculations. Dean explained that he could *see* the steps, so writing them down was pointless. Nevertheless, he felt stupid throughout much of his schooling on Long Island, especially compared to his brother Bart, two and a half years older, who excelled at academics and would be named a Westinghouse scholar as a high school senior. (He is now a distinguished pediatric oncologist.)

At the start of high school, weighing in at 105 pounds, Dean went out for football. He quickly abandoned that experiment and joined the wrestling team instead. Academically he remained bored, with marginal grades. When his friends got 90s and 100s on their math tests, he would retort that that was easy and challenge them to do what he did—purposefully get a 57. He swears this is true.

Then he discovered primary scientific texts such as Newton's *Principia* and Galileo's works. Vistas opened. He calls these pioneering scientists his real teachers. They would inspire many of his later inventions.

By the middle of high school Dean was playing around with the latest developments in electronics, powerful semiconductors and solid-state supertransistors called SCRs and Triax. He made a light box that could be plugged into a stereo so the lights pulsed with the music, and began putting on shows for friends in his parents' basement.

Soon after turning sixteen and getting a driver's license, he took a summer job working for a man who designed slide shows and wanted Dean to build cabinets for his projectors. It was mindless work, but it offered the excitement of driving into Manhattan, where one of the man's clients was the Museum of Natural History. Dean had a pass that let him into the museum's restricted underground garage.

He quit after a few weeks, bored. But he had a plan. He had noticed that the lighting system in the museum's Hayden Planetarium was old and bulky, so one day he drove into New York and waited to see the museum's chairman. When he was finally admitted, he told the chairman that he wanted to upgrade the museum's lighting system using state-of-the-art transistors and semiconductors. The chairman saw a cocky, scrawny, sixteen-year-old kid. He blew him off.

That provoked Dean. He took the money made from building cabinets, $80 or so, and spent it on parts at Radio Shack. For the next couple of weeks he worked day and night in his basement, designing a light show. Then he used his pass to enter the museum and wired his box into the planetarium's light system. Just like that. He found the chairman and gave him the news. You've done *what*? said the chairman.

Before he could be thrown out, Dean convinced the chairman to come take a look. When Dean flipped the switch, the rotunda burst into illumination. The chairman looked around slowly, then invited Dean to his office and asked a question: How much would this system cost the museum?

Dean was a kid. It seemed to him that his whole future depended on his answer, because he had quit his summer job and risked all his earnings. Working in the basement, he had dreamed about pocketing the immense sum of $1,000. So he swallowed hard and, characteristically, asked for twice that much.

The chairman walked around his desk. On one condition, he said: Dean had to do the same thing for the other three museums under the chairman's care. Four museums for $2,000, thought Dean, deflated. Not much profit. But before he could say OK, the chairman finished the math: The fee would be $8,000. Dean's gamble had paid off.

By the time he graduated from high school, he was selling light boxes to local rock bands and building customized audiovisual presentations that synchronized multiple slide projectors. He started college at Worcester Polytechnic Institute in Massachusetts, but ignored requirements and disregarded grades. Instead, he sat in on classes that interested him and chatted with professors about physics, engineering, and his heroes: Galileo, Archimedes, Newton, and Einstein. Besides, he didn't have much time for academic drudgery. He was driving home every weekend to work on his basement business, which he called Independent Prototype. He was making about $60,000 a year, good money in the early seventies, and exceptional for a college student. He sank it all back into the business, moving his mother's washer and dryer upstairs to make room in the basement for a lathe, a band saw, a milling machine, and an oscilloscope.

A man named William Murphy, the founder of a medical company called Cordis, had admired the show at the Hayden. Murphy wanted to hire whoever had done it to design an audiovisual presentation that would introduce Cordis's new programmable pacemaker at a big trade show in California. Someone from Cordis left a message on Independent Prototype's answering machine. Dean agreed to a fee for a twenty-four-screen show. Cordis sent a team to explain their needs. Dean arranged to meet them in New York City, figuring it was best to avoid the basement at this point in the relationship.

His age rattled them. "But I convinced them in the usual way that I could do the job," he recalled, "before they could realize that the entire company was one employee and he was in high school" (Dean's younger brother Mitch). After completing the project, Dean invited the Cordis people to come see it. Once they got over the shock of descending into a suburban basement to preview their expensive show, they were thrilled.

Some weeks later, Dean got a frantic call from California. The trade show opened in two days but the presentation wasn't working. Dean was sure he could fix anything wrong and agreed to fly to Anaheim that evening. The Cordis people who picked him up complained to him that Independent Prototype should have sent an engineer, not a kid.

Dean crawled under the stage and saw that someone had installed the transformer backwards, reversing the connections and melting the power supply. He pointed this out to the Cordis people, noted that it wasn't his fault, and wished them luck. Hysteria. So he called Long Island, woke up Mitch, and told him to start building a new power board.

Dean caught the red-eye home, finished and tested the board that afternoon, and caught the same flight back to California that he had taken the evening before. The Cordis people were frantic, certain that this kid could not possibly salvage the presentation before tomorrow morning's opening. Dean crawled under the stage and started soldering in the new board—ten connections, twenty connections, thirty. By the time he finished it was 5 A.M. and he was alone in the hall. He hadn't slept in days and decided to grab a few winks right where he lay, on the soft carpet beneath the stage.

When he woke up he saw feet everywhere. He crawled out into a pod of Cordis people, who besieged him with questions. A tall, dignified man approached, listened, and said that the solution was to turn on the presentation and see what happened.

It ran perfectly. That tall man, Cordis founder William Murphy, invited Dean to breakfast. Between bites, Dean informed Murphy that Cordis employed a lot of dunces. Murphy listened, amused and intrigued by this cocky young man who had crossed the country three times in twenty-four hours to fix his presentation. When Murphy traveled to New York a few weeks later, he surprised Dean by coming to visit. Dean showed him his basement workshop and his newest project, a device suggested by his brother Bart, a medical student at Harvard. Murphy examined the invention. It confirmed his hunch about Dean.

In those days, nurses had to monitor hospital IVs constantly to make sure they were working correctly. Bart had encouraged Dean to design something that would free nurses for other duties by automatically delivering precise doses of drugs at precise intervals. If the device was portable, so much the better, because that would liberate people who were otherwise healthy but tethered to hospitals and clinics by IVs.

Dean thought about the problem and then bought some inexpensive parts straight off the shelf—timers, counters, motors, batteries, circuit boards. He taught himself how to mill components out of aluminum. When he finished, he had invented the first drug-infusion pump. It was simple, portable, and precise. Bart showed it to people at Harvard Medical School, who suggested a few modifications. When the *New England Journal of Medicine* did a story about the pump, doctors called from all over the world. The National Institutes of Health ordered a hundred pumps at $2,000 each.

It was Dean's first big break. His younger brother Mitch and Mitch's friends became assemblers. His mother tested the circuit boards and kept the books. His father drew the illustrations for the manual. Dean named his new company AutoSyringe Inc. He was twenty years old.

To meet the demand for pumps, he needed more equipment, but some of it was too big to fit down the basement steps. So he sent his parents on a vacation and had a construction crew jack up the house, break through the foundation with bulldozers, and expand the basement deep into the backyard. A crane deposited large pieces of equipment, including lathes, a grinder, and a full-sized Bridgeport milling machine. Then the house was lowered back onto its foundation. When his parents returned, they were delighted with their new patio, the aboveground evidence of the underground expansion.

Dean also expanded his workforce, raiding Worcester Polytechnic for professors and students to help him improve his original design. He and his engineers created infusion pumps for chemotherapy, neonatology, and labor. Next, they designed the first portable insulin pump, which delivered a precise flow of insulin and thus freed diabetics from the debilitating fluctuations of their disease.

AutoSyringe eventually outgrew its basement headquarters and moved into an industrial space on Long Island. Dean continued to live with his parents. After paying tuition at Worcester Polytechnic for five years, he finally dropped out.

In 1979, fed up with New York's taxes and bureaucracy, he saw a New Hampshire license plate with the motto "Live Free or Die" and thought, "That's the place for me." New Hampshire had other attractions too: cheap real estate, no state income tax, hands-off government, and an underemployed workforce. Dean relocated AutoSyringe and convinced twenty employees to move with him. For the first time, he got his own apartment.

His pumps had become standard hospital equipment, and he couldn't keep up with the demand. His one hundred employees weren't enough. Competitors were pushing into the market. He faced a choice: Expand into serious manufacturing or sell out to a bigger company.

The drudgery of manufacturing didn't interest Dean. William Murphy, from Cordis, had introduced him to people at Baxter Healthcare, the medical conglomerate. In 1982 Dean sold AutoSyringe to Baxter, reputedly for $30 million. The handful of people who had invested to fund AutoSyringe received a substantial return.

At age thirty-one he was an unemployed multimillionaire. He bought a pleasant house on several wooded acres in Bedford, New Hampshire, outside Manchester. When his parents visited, he drove them to the house next door and said, "How do you like your new home?" They moved from Long Island.

He rewarded himself with a small plane and a helicopter, and learned to fly both. He especially loved the helicopter. "I remember lying in bed as a kid," he said later, "thinking how neat it would be to just hang there and defy gravity, and to be in a machine as agile as I am. Then I got older and saw helicopters."

He called the helicopter his magic carpet, and decided to buy a magic kingdom to go with it—an island in Long Island Sound called North Dumpling. Its three acres were landscaped with flowers, willows, mountain laurel, and Japanese black pines. There was a lighthouse in addition to a roomy modern house.

North Dumpling brought out the whimsical side of Dean's habitual insubordination. On the island he called himself Lord Dumpling. He drafted a North Dumpling Constitution and appointed friends and family to the Cabinet, which included ministers of Brunch and Nepotism. Ben Cohen and Jerry Greenfield of Ben & Jerry's served as Joint Ministers of Ice Cream, a vital office considering Dean's appetite for the confection. "We went off gold and went on the ice cream standard," he once said. "As long as we keep it below 32 degrees, our currency is rock solid." The kingdom's postage featured a turtle. Its currency bills had the value of pi. (Dean kept one of the bills in his wallet and sometimes amused himself by offering it as legal tender on the mainland.) The national anthem, sung to the tune of "America the Beautiful," expressed some of Lord Dumpling's deepest convictions:

> *North Dum-pling, North Dum-pling,*
> *Keep lawyers far from thee!*
> *And MBAs, and bureaucrats,*
> *That we may all be free!*

When the state of New York objected to the wind turbine he built to provide power on the island, Lord Dumpling threatened to secede from the United States. He decreed a territorial limit of 200 inches and signed a nonaggression, mutual-defense pact with President George Bush (George W.'s father), whom he knew through John Sununu, former governor of New Hampshire and President Bush's chief of staff. Dean hung an aerial photo of North Dumpling in his office with the caption, "The Only 100% Science-Literate Society... America Could Learn a Lot From Its Neighbor."

Nevertheless, in an aggravating assertion of state's rights, New York insisted on sending an inspector to look at the turbine. Dean arranged for the local Coast Guard commander, whom he had befriended, to transport the bureaucrat to the island and play along with the idea of Dumpling independence. As the commander and crew approached the island, they startled the inspector by shouting, "Permission to enter international waters!" When they prepared to step ashore, Lord Dumpling asked them to leave their guns on board, because visitors to Dumpling could bare legs but couldn't bear arms. The Coast Guard contingent also submitted to the Lord's visa process, in which an official paper was stamped Dumpling Bozo or Dumpling Bimbo. The inspector wasn't amused. After examining the turbine, he pointed to the lines going into the house and said he needed to see where they ended. Lord Dumpling regretfully informed him that that was a

Dumpling state secret. The inspector left in a huff, but Lord Dumpling had no more trouble about the turbine.

Dean visited the island only once or twice a year, and never for more than a day or two. Leisure made him jumpy. Nevertheless, he kept the island. "The psychic security of knowing that as everything goes to hell I have a place where nobody can bother me," he once said, "makes it worth it."

Dean didn't spend all his new money on expensive whimsies. He also bought a long stretch of Manchester's old Amoskeag Millyard along the Merrimack River—more than 500,000 square feet of brick factory buildings. At the end of the nineteenth century, these sixty-four textile mills comprised the world's largest industrial complex and were the pumping heart of Manchester. By the early 1980s, they had deteriorated into an eyesore. But they were also cheap and structurally sound, a bargain for someone with vision. To Dean they symbolized a time when American engineering and technology had led the country to greatness and changed the world. That was the spirit he wanted for his new R&D company, which he named after himself: DEKA (DEan KAmen).

While the first mill buildings were renovated, DEKA operated out of Dean's house, where Dean had added a full machine shop just off the living area. He liked to relax late at night by fabricating parts. The rehabbed mill buildings needed to be heated and cooled, but Dean considered the available systems deplorable, so he designed his own, controlled by personal computers. He patented his innovations and started a spin-off company, Teletrol, that installed climate-control systems for large commercial and industrial buildings, including the Sydney Opera House in Australia, Disney's Parisian headquarters, and NASA's Mission Control in Houston.

DEKA eventually occupied 70,000 square feet in two of the mill buildings. As Dean rehabbed more buildings, new tenants moved in, reviving this formerly run-down area on the edge of downtown. Dean's real estate holdings alone would have assured his continued wealth.

Commuting to work in his Enstrom helicopter, he couldn't help noticing a few ways to improve the machine, so he bought Enstrom and soon owned a number of helicopter patents.

All this economic activity and dramatic heli-commuting made a big splash in the small pond of New Hampshire. It helped that Dean was young, brash, and quotable. He dated several women, including a former Miss New Hampshire, before meeting a waitress named K. C. Connors in January 1994. Fifteen years younger than Dean, K. C. soon moved in with him and went to work for FIRST. Warm, outgoing, and devoted to Dean, she would live with him for nearly eight years.

By the mid-1990s Dean's local celebrity was such that Manchester's newspaper, the *Union Leader,* ran a story when he shaved his beard. He became friends with some of the local movers and shakers, including the governor, John Sununu, a former engineer, who would give Dean entrée into the first Bush White House.

Not everyone found Dean charming. One of his wealthy neighbors in Bedford complained about the helicopter's noise. When the town supported Dean, the homeowner sued. The case went all the way to the state Supreme Court. By that point Dean had fired two law firms because they were willing to settle for a variance, which offended his sense of rightness. He argued his own case and won.

His helicopter also caused a dispute between two bureaucracies, the sort of battle that Dean entered with grim glee. Manchester's municipal regulations required a 30-inch fence around all helipads, but the Federal Aviation Authority's regulations said that anything above 10 inches would be illegal on DEKA's roof. He managed to obey *and* flout the rules by buying an orange mesh fence, 30 inches high, and laying it flat around his rooftop helipad. The city inspector objected. "You said I needed a 30-inch fence," replied Dean. "You didn't say it had to be standing up." He won.

Meanwhile, for several years Dean and his engineers had been working on another revolutionary medical device. In 1987 Baxter Healthcare had asked Dean to improve its kidney dialysis machine, used by diabetics who needed their blood cleansed. The existing machine was noisy, expensive, heavy (180 pounds), and as bulky as a freezer. Its gravity-driven technology hadn't changed much since the 1950s. Technicians had to be trained to operate it and to route tubing through its convoluted system of valves. For all these reasons, the machines were located in hospitals and clinics, forcing patients to travel to them. The cleansing procedure took several hours. Some diabetics required the treatment twice a week. Everything about it was inconvenient and antiquated.

Dean agreed to improve the machine, but he didn't really mean it. He made a half-hearted show of upgrading the mechanical valving, but quickly dropped that line of inquiry. Improvements didn't interest him; transformation did. "Don't solve the *solution*," he liked to say, "solve the *problem*." Any engineering firm could twiddle a revision. But clients often balked at radical ideas, so Dean tended to keep them to himself until he could make them irresistible.

His drastic plans for the dialysis project alarmed the only two M.B.A.'s at DEKA, a reaction that confirmed Dean's view of their degrees. He told them that if J. P. Morgan had said to them, "I want to build a railroad to the West Coast," they would have advised against it as too capital intensive, with an uncertain return, because the railroad would be going into nowhere. Morgan's response to such sensible M.B.A. advice, added Dean, would have been, "Morons! I *know* there's nothing out there. That's *why* I want to build the railroad!"

Dean approached new projects the same way, as if he were heading into wide-open territory where anything was possible and history didn't block the view. He always imagined that he was tackling a problem for the first time, and urged his engineers not to limit themselves to what they thought was feasible. Nothing was too wild to consider. The best ideas always seemed impossible at first, or at least improbable.

Dean hired only the best engineers, but mere brains weren't enough. He wanted engineers who thought it was fun to jump off a cliff and design an ingenious new parachute on their way down. Sometimes they crashed, but Dean didn't hold that against them if they learned something valuable en route. Engineers without the nerve to jump, and to keep jumping, didn't last long at DEKA. That made the company an exhilarating place to work, for a certain kind of engineer. Thomas Edison once remarked that no experiments are useless. Dean agreed. He was displeased if he and his engineers weren't frequently failing in preposterous ways, because impressive failures signified impressive aspirations. Puny failures sprouted from puny dreams.

So Dean encouraged constant technological audacity. "You gotta kiss a lot of frogs," he liked to say, "before you find a prince." He often condensed this phrase into a gerund—frog-kissing—and pushed his engineers to pucker up and experiment. This insouciance about innovation permeated DEKA and took the edge off failure.

For Dean, frog-kissing meant looking for ways to combine the laws of physics with the latest technologies. "History is a great teacher," he said, "but progress depends on disproving history." He liked to imagine how his heroes, the discoverers of the laws of physics, would attack a problem.

In the case of the dialysis machine, his musings took him back to Boyle's Law and Gay-Lussac's Law, now-obscure discoveries from the seventeenth and eighteenth centuries about the properties of gases. He explained the physics to his engineers and speculated that these principles could be combined with computer-controlled pneumatics and other modern technologies to create something altogether new for treating dialysis.

It took five years to turn that insight into a product. In 1993 Baxter introduced its HomeChoice dialysis machine. It weighed just 22 pounds, was affordable, and could fit under an airline seat. Because it used air pressure to regulate flow, it ran so quietly that people could dialyze themselves while they slept. It was easy to operate, too. The patient popped in a disposable cassette, attached bags of dialysis fluids to some tubes, and pressed the On button. People needing the treatment were no longer chained to hospitals. *Design News* magazine named the machine the best medical device of the year. The royalties from it became the main source of Dean's personal fortune, and hence funded DEKA's freelance inventions.

The dialysis machine fit the pattern that Dean had been following since his days designing light boxes and infusion pumps: Combine inexpensive new technologies in unexpected ways to do unforeseen things. "I don't have to invent *anything*," he once said. "It's out there somewhere if I can just find it and integrate it." He added, "Inventing is frustrating, it's dangerous, it's expensive, and inventors should avoid it whenever possible. Be a systems integrator." Innovation, he said, was "the art of concealing your sources."

The dialysis machine also started a pattern that Fred and Ginger would repeat: a flash of inspiration by Dean, followed by years of work by his engineers to make it real. The process brought to mind Thoreau's words in *Walden*: "If you have built castles in the air, your work need not be lost; that is where they should be. Now put the foundations under them."

"Yes," said Dean, nodding when he heard the quote. "That's what we do here."

While DEKA was still working on these projects, Dean was disturbed one day while watching a man struggle to get his wheelchair over a curb at the local mall. Inside, he saw the man straining to reach over a high counter for an ice cream cone.

The episode nettled him. We can put people on the moon and travel to the depths of the ocean, he mused, but we can't get a wheelchair over a curb? The more he thought about the limitations faced by people in wheelchairs—no eye-level conversations, no trips to the beach or anywhere else without sidewalks, no reaching the top shelf at the grocery, no possibility of defeating the absolute barrier of stairs—the more he became offended as an engineer and as a human being. "I'll fix it," he decided, certain it would be simple.

He collected all the wheelchair patents from the last hundred years. The ingenuity in the tall stack surprised him. He suspected that the solution would be a climbing mechanism. So had lots of people before him. They had tried movable legs, flexible arms, arms that grabbed. He admired the ideas and the elegant drawings, but studying them revealed their flaws.

He kept chewing on the question. As always, he tried to see the problem afresh, as if it had never been addressed. Maybe a wheelchair was the wrong solution. First, he thought, identify the basic issues: If I'm disabled, I don't want to be stopped by obstacles, including stairs, and I want to be able to move at eye level with everyone else. He considered and rejected the idea of four legs, because two would have to lift at once, causing instability. That's why insects have six legs, he decided, because they can always have three legs down, creating a pivot. But six legs would be a technological nightmare.

It had looked like such a simple problem, but he couldn't think of a solution. The puzzle often kept him awake at night. He put a couple of engineers on it, but it stumped them, too. Two years passed, two years of princeless frog-kissing. Dean was ready to give up.

Archimedes had his eureka moment in the bathtub. Dean had his as he stepped out of the shower and slipped on some wet tiles. His legs started to hydroplane out from under him, so he instinctively flung back his arms and wheeled them in reverse to recover his balance. Standing there dripping, it hit him: That's *it*! I've *solved* it!

The missing piece was balance. Dean told a few engineers to start kissing frogs that could maintain equilibrium. By July 1992, they had built Rev 0 (rev is engineering shorthand for revision). They used off-the-shelf parts: amps, two printer motors that cost $10 each at a local surplus shop, and a gyroscopic tilt sensor called an inclinometer (originally developed for gun turrets on battleships), which was, at about $100, the machine's most expensive part.

It was a sorry-looking contraption, the size and height of a runty end table. Batteries and bare wires were taped together and held in place by Velcro. Ragged foam cushioned the worst steel edges. Chains formed the drive train. Thick wires connected all this to a separate control board operated with a joystick, which made the thingamajig's little wheels turn. It looked poor and half-naked, but it achieved its purpose as a "proof of concept": It balanced in place, trembling, in a state that engineers call "statically stable." If someone pushed it, it regained its original position.

Essentially the team had taken something that wanted to fall down—imagine a table with two wheels instead of four legs—and forced it to keep standing. Engineering textbooks call this "balancing an inverted pendulum using an automatic controller." The exercise entails measuring the pitch angle (the tilt), the pitch rate (changes in the pitch angle), the wheel position, and the wheel speed (changes in the wheel position). The task was to read all these variables, calculate the adjustments necessary to maintain balance, and instruct a power source to make those adjustments, called gains.

Rev 0's jiggly equilibrium tickled the team. You could put a bowl of pretzels on it and move it around the lab. But what about a seated person? Engineers had known the formula for inverted pendulums for decades. Dean's plans were audacious because no one had imagined putting someone *on* something like this and *moving* in balance, much less climbing stairs with it. Imagining such a thing was brash; achieving it, improbable. Most difficult of all, how could you make such a machine safe for disabled people who couldn't hop off if something went wrong? The team was having too much fun experimenting to worry about the grueling terrain ahead.

Dean wanted more engineers on the project. One of his old professors at Worcester Polytechnic recommended an instructor there named Kurt Heinzmann. Kurt was the son and grandson of engineers from Connecticut, but as a boy he didn't care for school, preferring to collect animal bones or tinker with generators and motors. He graduated from high school with a C– average, certain he was finished with school forever.

After working in a local brickyard for a while, Kurt drifted up to Maine as a farm laborer. Denver was next: construction worker, car mechanic, line cook in a nightclub. After two years at the club he noticed that he was indulging in too much free scotch, so he quit to work as a gandy dancer for the railroad on the western side of the Rockies. Then he returned home to the brickyard, where he stayed for four years. His skill with the machines lifted him to the position of plant manager. But he was restless. He started dabbling in courses at a community college and decided that maybe he'd been wrong about school. He ended up with a master's in engineering from Worcester Polytechnic, where he became a part-time instructor.

Kurt was a perfect fit for DEKA—nonconformist, blunt-spoken, curious, inventive, happy to get dirty, inexhaustible. He was skinny and broad-shouldered, with blazing eyes. His time in Colorado had shaped his wardrobe, which consisted of Western shirts with snap buttons, pointy cowboy boots, and tight faded jeans. He looked like a bronco buster, and in a way that's what he became at DEKA.

He jumped into Rev 1. At this point its wheels lifted for climbing stairs, which it managed to do on good days with help from swiveling pigeon-toed feet. These platform feet, propelled by actuators, had been designed for the space program and cost $15,000 each. But Rev 1 was unstable on stairs—that is, it constantly crashed.

In December 1992, soon after Kurt joined the team, Dean decided to recruit another frog-kisser who would become vital to the project. Dean's right-hand man at DEKA was an engineer named Mike Ambrogi who had graduated at the top of his class at MIT. Mike's younger brother had followed him through MIT, then taken a job at Hughes Aircraft in California. His name was Robert, but everyone called him Benge (pronounced Benji). After quizzing Mike about

Benge's talents and personality, Dean decided that another Ambrogi could only help DEKA. He asked Mike to call and sell his brother on the idea. When Benge said he was happy where he was, Dean grabbed the phone: "I want to show you what we're working on."

Flattered, Benge agreed to visit DEKA. His family still lived in New England, so why not take a free holiday trip? He had no intention of moving from Redondo Beach, a small town on the Pacific. He liked to surf, liked knowing most of the people in his neighborhood, and liked working on a classified infrared telescope. His brother knew it would take a lot to lure him out of California.

Benge arrived in Manchester early on a dismal morning in December—bare trees, slushy roads, gray skies, freezing air. "One of those days that make New England look *bad,*" he recalled. He turned onto DEKA's street, lined with old mill buildings, most of them dilapidated. "I don't want *anything* to do with this," he thought.

At DEKA he met Dean but at first didn't know who he was because of his casual demeanor. An engineer gave him a tour. Benge liked DEKA's intense yet informal atmosphere. It impressed him that an engineer could go straight to the machine shop and get a part made. At Hughes the process took forever. He spent Saturday at Dean's house, where Dean sold him hard on DEKA's freewheeling philosophy and exciting projects. As a finale, Dean showed him Rev 1. Benge thought it was a cool curiosity, but no big deal.

He was happy to get home to California. "But then my wife and I were having sushi and—," he paused, still sounding puzzled a decade later. "I really don't know how it happened, but—well, basically it was the *aura of Dean,* that unquantifiable *something,*" he said, shaping the air with his lively hands.

He moved to New Hampshire in April but steered clear of the "transporter," as Fred was called then. He thought DEKA's future, and his, lay in the pumping technology developed for the dialysis machine. But Kurt Heinzmann kept stopping by his desk to pick his brain and talk about the transporter. They liked each other. Unlike Kurt, Benge was short and compact. But both of them radiated confidence and liked to shoot from the hip, and both bristled with impatient energy. Benge walked so fast he nearly trotted, and people always knew when Kurt was coming because the heels of his cowboy boots rat-a-tat-tatted like a nail gun.

Soon after Benge arrived, Dean showed the transporter to some-one from the outside who told Dean it was Nobel Prize material. This outsider warned Dean to protect the revolutionary technology from theft by taking the project undercover. This aggravated Dean's natural condition as an inventor: paranoia. He put the transporter behind locked doors on the second floor. Before then, everyone at DEKA had been encouraged to poke their noses into anything going on. Now Dean's anxiety and ego had separated DEKA into two groups: those who knew the secret and those who didn't.

When Dean urged Benge to move downstairs with the secret project, the younger Ambrogi became the team's fifth member. For weeks the team tried different approaches to mobility. Hydraulics, ro-botics, wheel spokes that collapsed. They even flirted with the idea of a hovercraft. Dean encouraged them to try their craziest ideas. It was fun, but frustrating. Nothing but frogs. Even the designs that managed to climb a few stairs were unstable. "We had lots of systems, all bad," re-called Benge. They developed new respect for the complicated me-chanics behind the act of walking.

Finally one day Dean came down and said, "This isn't working. Start over."

He suggested a rack-and-pinion arrangement like the one used by the cog railway that climbed nearby Mount Washington. In Dean's ver-sion, the transporter would lay down its tread whenever it needed to climb stairs.

Benge found the idea amusing. Kurt hated it, which spurred him to look for alternatives. He designed something that resembled a sim-ple gear, which he called a *cluster*. A machine with a cluster on each side could climb stairs leg over leg, a refinement of Dean's cog ramp. In Kurt's design, the stairs functioned as the rack and the legs were the en-gaging gears. But the clusters, like the other designs, were unstable. More fun, more failure.

One day at a brainstorming meeting someone wondered out loud—no one can remember who it was—whether they should recon-sider their requirement that the machine had to be statically stable on its wheels. Could something unstable work?

The right question is always an embryonic insight. The team quickly realized that instability was worth exploring. The act of walking was inherently unstable yet worked beautifully. So far, their designs had mimicked the way walking looked (moving legs, articulating joints), but not the way it actually functioned. The technological solution wasn't to replace a disabled person's legs, but to mimic a healthy body's ability to balance despite an unstable foundation on two small pivots (feet). It was the logical extension of Dean's discovery on the slippery tiles.

When we climb stairs, Dean and the engineers realized, we aren't statically stable, we are *dynamically* stable. That is, we are slightly unstable at any given moment, but we use our internal gyros to compensate automatically. We defeat instability every time we take a step. The act of walking entails a series of controlled falls, instantaneously prevented by our dynamic balance. Balance makes it possible to climb stairs, or lean without falling, or regain equilibrium after slipping on wet tiles. Yes, the team realized, dynamic balance was the answer. The new task was to put that insight into a machine.

Our dynamic stability depends on speed—specifically, reaction time. The team was already using microprocessors that could handle ten thousand instructions per second, as well as advanced gyroscopic technology that had been designed for handheld video cameras (steady cams). The goal was to combine dynamic stability with stair climbing.

Kurt wanted to resurrect his old idea of "clusters" on a pivot, but this time he wanted to put a small wheel on each of the gear's three legs. The others didn't pay much attention. One Saturday in the fall of 1993, Benge went to Dean's to play with him in his machine shop. Why not fool around with the cluster idea? They cut two three-legged gears out of structural foam with a scroll saw, then attached a small wheel to each leg. They began rolling the components up and down the stairs by hand, end over end. As the forward wheel hit the back of a stair and stopped, momentum pushed the following wheel up and over to the next step, leaving the third wheel in position for the next revolution.

Maybe there was something to Kurt's idea. They decided to try it in metal. Benge designed the part on a computer, and they machined two of them. The components weighed 1.3 pounds each, but were strong. You could put a Buick on this, said Dean, imagining the weight of a human payload. He was getting more and more excited.

For three consecutive nights he and Benge worked at Dean's shop until 2:00 or 3:00 A.M., building a cluster machine out of metal. When they needed to weld a base onto the plywood seat with aluminum, which neither of them knew how to do, Dean remembered that a DEKA employee could weld aluminum and had a home shop. They flew to the employee's house in Dean's helicopter, landed in the yard, and put him to work. They finished the machine late on the third night, but had to wait until the software engineers plugged in some computerized intelligence to see how it worked.

The team called the machine No. 5, after the personable robot in the movie *Short Circuit*. They experimented with the clusters. One day, goofing around, they rotated them in opposite directions, which made the machine dip and jig in a crazy rumba that led someone to remark that it danced like Fred Astaire.

"No," said Dean. "Fred *Up*stairs." They began calling the project Fred.

Operating on the principle that walking is a series of controlled falls, they designed a machine that would begin to climb stairs while balancing. As the clustered wheels revolved, the software would kill the balance command at a certain moment, causing the machine to fall forward onto the next step. Not a reassuring solution.

One day Kurt said, "What if, instead of balancing on the wheels, we balance on the clusters?" In other words, tell the inclinometer to listen to the clusters, not the wheel. That way, the clusters would control the balance. They changed the software to make this possible. Kurt sat on the machine at the base of a stairwell, facing away from the stairs. He grabbed the handrail and leaned back.

The machine climbed right up. Even Kurt was dumbfounded. "It was like setting off a rocket for the first time," he recalled. "That was a major breakthrough. Suddenly we had a machine that could fall *up* stairs. It didn't have to know whether it was supposed to go up or down, it just responded to the way you leaned. If you leaned toward the top of the stairs, it would climb. If you leaned toward the bottom, it would go down."

Clustered wheels wasn't a new idea. The innovation was that the team, principally Kurt, devised a way to control the clusters by making two simple changes to the formula for balancing an inverted pendulum. He changed the gains and, more important, instead of measuring wheel torque, he measured cluster torque.

By December 1993, seven or eight people were working on Fred full time. The number grew steadily, including two engineers who would later form part of Ginger's core team: John Morrell, a controls engineer, and Kurt's older brother, John David, called J. D. Like Kurt, J. D. was a tinkerer and dabbler who could do all kinds of engineering.

The team built a series of increasingly sophisticated machines with names such as R2-D2, Baryshnikov, Nureyev, Clyde, George, and Grumpy. They tried to do things cheaply, with parts close at hand. For various models of the seat they used plywood, an old car seat, a bass-fishing seat from Wal-Mart, and big biker seats from a local motorcycle shop. They also used off-the-shelf mechanical parts, not just because they were less expensive (Dean was funding Fred out of his own pocket), but because custom parts could make the machine unfeasible to manufacture in the future.

They got banged around by the usual kinks, mistakes, and failures. "It took three months each to build Baryshnikov and R2-D2," recalled Benge, "and then they were like horses—it took another six months to break 'em."

A video from that time captures the spirit of the lab, with Benge zinging around on R2-D2 in a horned Viking helmet. In another clip, the machine goes so fast that it tilts too far forward, forcing Benge to hop out and jog ahead, Fred in pursuit. There's a segment of Benge and Kurt racing. Kurt wins by a nose, but then his machine goes into an epileptic fit. The same tape shows someone lying on the ground right after a crash. Kurt, looking concerned, bends over for a closer look—at the machine, not the human victim. "It looked like the clusters kind of slipped out from under you," he muses. Another clip shows Fred ejecting an engineer and then scooting off-camera. As the engineer sits there scowling, Fred zips back into the picture and rams him in the back. Once, Fred suddenly switched into stair mode and tried to climb up somebody's pants. The software engineers stayed busy rewriting code, trying to smooth the machine out and make it less psychotic.

DEKA's payroll had increased from a dozen to five dozen employees. Most of them worked on paying projects, but the Fred team kept growing, too. Two years of research and development had cost Dean a few million dollars. The good news was that he and his engineers had solved the basic technological problems and proven the invention's feasibility through prototypes. The bad news was that it would take millions more to get the machine to the marketplace. Fred needed a sugar daddy.

The project stayed behind locked doors, but Dean and the team began giving demos to important outsiders to get feedback and to angle for funding. Benge called these visits "the parade of CEOs." A video of one such demo shows a dignified executive type sitting on Fred as smoke streams from the base. Someone off-camera says, "Sir, maybe you should get off." Baryshnikov had a habit of launching itself into walls just before demos. The team always hovered nearby at these performances, in case the unruly beast tried to assault some potential patron.

A number of companies were interested in Fred. Things progressed so far with one electronics conglomerate that Dean told the team to get some desks for the company's reps. But he balked at the company's terms and kept procrastinating. When the company gave him a deadline, he made a last-ditch call to Dr. Robert Gussin, senior technology officer at Johnson & Johnson. This was a Thursday in late 1994. The two had gotten to know each other when DEKA designed an intravascular stent for Johnson & Johnson. (Vice-president Dick Cheney has one in an artery.) Dean asked Gussin to come see something at DEKA. Gussin tried to beg off. He was expecting guests the next day at his Long Island beach house.

You *have* to come, said Dean. It's the best thing I've ever done.

That got Gussin's attention. He flew to Manchester on Saturday. Dean and some of the team showed him Rev 0. Gussin thought it looked like an ugly card table on wheels and asked himself, "*Why* am I here?"

Then he realized that the table was *standing* there, balancing. He couldn't believe it. Dean showed him the other prototypes, which could climb stairs. Gussin was astonished. Dean suggested that Johnson & Johnson might want to use this technology to get back into the wheelchair business. But he needed an answer within a week. Instead of asking for his usual hefty royalty, Dean offered Johnson & Johnson a different deal. He figured it would cost tens of millions more to get the machine to market. If Johnson & Johnson would put up the money, Dean would accept a smaller royalty. All he wanted in return were the rights to all nonmedical applications.

The prospect excited Gussin. He sent a memo about it to Ralph Larsen, Johnson & Johnson's chairman and CEO. They had a good rapport and Larsen usually responded quickly. When Gussin hadn't heard anything by Wednesday, he left another message. He left a third message on Thursday. On Friday he told Larsen's assistant he had to see him.

He's busy until 4:00, said the assistant.

OK, said Gussin, I'll see him at 4:00.

No, said the assistant, because he's leaving on the helicopter for a meeting in New York.

OK, persisted Gussin, I'll talk to him at the helipad before he leaves.

He might not have time, said the assistant.

I'll ride with him in the helicopter, said Gussin.

Might not be any room, said the assistant.

Look, said Gussin, one way or another I'm talking to him today.

He hung up. A minute later the phone rang.

Bob, said Ralph Larsen, we are *not* getting back into the wheelchair business.

This isn't just a wheelchair, said Gussin. You *have* to see this thing.

I don't care what it is, said Larsen. It's hardware and I'm not interested.

If you don't do this, said Gussin, you'll be making the *biggest mistake of your career.*

That got Larsen's attention. He sighed and agreed to look at it. In Manchester he listened to Dean's relentless spiel and watched Fred dance, then turned to Dean and said, "It's a deal." Johnson & Johnson was back in the wheelchair business and Fred had a sugar daddy.

Dean threw a celebratory dinner for the team and their spouses at the Bedford Village Inn, the poshest place in the area. The Christmas decorations were up and the mood was festive, until Dean mentioned that Johnson & Johnson was insisting on a three-year deadline. The team nearly choked.

Benge spoke up. He meant to ask, "What happens if we don't make it?" but what slipped out was "*when* we don't make it?"

Dean launched a diatribe about meeting their obligations, spending Johnson & Johnson's money wisely, and changing the world for all the people who would benefit from a balancing wheelchair. He talked nonstop for an hour and a half, not unusual when the spirit possessed him. People kept looking at their watches and at each other, feeling restless or anxious about their babysitters, but they didn't move. As always, Dean's vision pinned them.

After Dean signed the papers with Johnson & Johnson, the company sent a group of its top executives to look at Fred. Benge, Kurt, and the others took a machine over to Dean's for a demo to impress the investors. Fred behaved itself, and afterwards the executives moved into the house for drinks.

The engineers were still outside with the machine when the seat weld broke and Fred went insane, jittering out of control. The executives, turning toward the commotion, saw Fred body-slam Benge into Dean's picture window. They watched in shock as their multimillion-dollar investment banged at the glass in a frenzy, as if trying to get at them.

For weeks afterward, Johnson & Johnson's people kept asking Dean about this incident in alarmed tones. Dean worried that the company might pull out of the project because of it. But the squall passed, and the episode gradually became another piece of amusing Fred lore. At one of DEKA's monthly meetings, Dean arranged for Benge to receive a citation: *DEKA Academy Award Nomination, Best Individual Scene: Rear Window.* In retrospect, their worst mistakes and failures always made them laugh the hardest, because they had survived them.

Sometime before Dean showed Fred to Johnson & Johnson, Kurt Heinzmann, who had excellent balance and was crazy, had started entertaining himself by standing on Rev 0's small platform and using the joystick to surf around the lab. For a while no one, including Dean, realized the implications of this larky feat. But then they did, and Dean's imagination leapt at the possibilities. That's why he had insisted on retaining the rights to all nonmedical applications of the technology.

He asked the team to package what they had learned about dynamic stability into a small, low platform with two wheels, something someone could stand on, with a vertical bar connected to a handlebar. They built it and hopped on. When you leaned forward, it moved ahead. The more you leaned, the faster it went. To stop, you leaned slightly backward. It was just like walking, but more fun. The team was too busy with Fred to give it much attention. For a while it was little more than a diversion. It was lighter and slighter than Fred Upstairs. They named it Ginger.

chapter two

A Dream with a Deadline

Summer 1999–October 1999

After Fred found a sugar daddy, Ginger became DEKA's newest unfunded orphan. In practical terms that meant working on Ginger after hours. The little machine's possibilities seemed to intrigue Kurt more than Fred's other engineers. He built another rev and named it Mary Ann, after the clean-cut friend of Ginger on *Gilligan's Island*. But this Mary Ann looked slovenly, with hanging wires, crude metalwork, and lots of black tape holding things together. Yet it ran and it balanced. Mary Ann became Kurt's lab toy.

That's where things stood with Ginger in the fall of 1996 when J. Douglas Field showed up for his first day of work. Doug was in his early thirties, tall and slender, with thinning brown hair and soulful eyes. He had the relaxed yet earnest manner of a hip young priest. When fully amused he didn't laugh out loud, but turned bright red with the contained explosion.

His mother had written in Doug's baby book that he obviously would grow up to do something with wheels. She was right. Even as a young boy he wanted to work in the transportation industry, and as an adult he raced bicycles and obsessed over cars.

After earning a degree in mechanical engineering, Doug joined Sikorsky Aircraft, but found the pace too slow. He yearned for action and innovation. So he took a position at Ford Motor Company in product development, which he hoped would be his dream job. Instead he found himself stuck in the slow lane again, making incremental improvements on a luxury car that was "only 2 percent better than last year's model and still 3 percent behind Japan."

Ford did pay for him to get simultaneous master's degrees in engineering and management from MIT and MIT's Sloan School. He gobbled up books about legendary engineering projects—the personal computer at Xerox's Palo Alto Research Center, the Stealth fighter and the U-2 at Lockheed's top-secret Skunk Works, the Macintosh at Apple. These stories made him ache to be part of something electrifying.

After five years of "crushing frustration" at Ford, he needed to break out. In his emotional way, he characterized his decision to leave the transportation industry as "traumatic." He moved to Johnson & Johnson. Interesting work, but he missed wheels. There he met Dean, who, with his sharp eye for engineering talent, mentioned to Doug that he had just the project for someone with his passions.

Doug had spent his career in large corporations. Working for a small R&D company would be a big change, so it took him a while to find the nerve to send Dean a letter. But a few days after doing so, Doug was in New Hampshire undergoing one of DEKA's grueling interviews. He did fine, but when he wasn't shown the secret project he would be working on (Fred), he balked at the job offer. Harley-Davidson was pressuring him to take a job too. The night before he had to choose, he woke up in a cold sweat with the clear conviction that even though he didn't know what he would be doing at DEKA, working on a $15,000 motorcycle would be trivial.

So he took Dean's offer, and on his first day at DEKA he saw Fred. "I had a huge panic," he recalled. "It had *wheels* and it was *new*. Oh my God! What if I *hadn't* come here, and then I saw it when it came out?" The idea was almost unbearable, which reveals his temperament and aspirations.

On that first day he saw something else that made his jaw drop: an engineer in cowboy boots and a Western shirt, holding a cup of coffee as he zipped down the hall on a strange little two-wheeled contraption called Mary Ann. Good-bye slow lane, hello autobahn. Doug went home exhilarated.

A balancing, stair-climbing wheelchair certainly was attractive work for an ambitious engineer, but lithesome Ginger was bound to steal Doug's heart. He soon had a crush on Mary Ann. It was a rough courtship. The machine slammed him into a door at 15 miles per hour, jerked him around, afflicted him with stitches and a foot-long rug burn. "She can really be a bitch sometimes," he said, smiling like a man in love.

In June 1997, Doug and two other Fred engineers began moonlighting at Dean's home machine shop, working on the next rev of Ginger. Doug still treasures their first drawings, done on napkins. They frequently stayed at Dean's until 4 A.M., machining parts and experimenting. Dean often joined them. On the night of final assembly for the new rev, tingling with anticipation, they discovered they didn't have the right bolts for the last few connections. They jumped into Dean's helicopter and flew to DEKA to pick some up.

The new rev had different tires, motors, and transmissions. It also had multiple personalities that changed abruptly, so they named it Sybil. It could reach 20 miles per hour. The consequences can be expressed in a nontechnical equation:

$$(\text{Speed} + \text{Volatility})\,\text{Unpredictability} = \frac{\text{Crashes}}{\text{Time}}$$

Doug loved all of Sybil's peculiarities and spent as much time with the machine as he could spare from his duties on Fred, where he had become chief engineer.

Near the end of 1998, when Doug had been at DEKA for two years, Dean decided to get serious about Ginger. He made Doug the new project's first employee and director of engineering, with responsibility for putting together a team. Doug was ecstatic. He and Ginger fit together like helical gears. This was the project he had always dreamed about: designing and developing a new paradigm—with wheels!—and then taking it to market. "It has taken me over," he said during the project's first year. "I owe Dean for giving me the greatest experience of my life." Many people in the project came to believe that Doug's impact on Ginger was bigger than Dean's. They have a strong case.

Most engineers concentrate on function and disregard form, but Doug cared fervently about aesthetics. As a respite from his dull jobs before DEKA, he had taken art and jewelry lessons and had studied design. He was determined to make Ginger not only technologically amazing but pleasing to the eye. His first outside hire wasn't an electronics whiz or a mechanical mastermind, but a young industrial designer, a creature hitherto unknown at DEKA.

His name was Scott Waters. Still in his twenties and rather short on experience, considering the scope of his task at Ginger, Scott was gifted and passionate and unafraid to break molds. Doug had learned from Dean to hire bright young prospects who didn't yet consider anything impossible. Like Doug, Scott was tall, with an easygoing personality that camouflaged an emotional nature. He would be tested both by the demands of designing a completely new product and by the indifference to aesthetics among many of the engineers.

Doug and Scott worked hard to formulate a "Ginger Design Philosophy." This one-page document, ceremonially signed by them on July 9, 1999, functioned as a manifesto. The team often referred to it as a reminder of first principles, a design Bill of Rights (and Wrongs). It was sprinkled with the inspirational quotations that Doug collected, many of which became Ginger mantras:

"Everything should be as simple as possible, but not simpler." (Albert Einstein)

"Good artists borrow. Great artists steal." (Picasso)

"Real artists ship." (Steve Jobs)

The manifesto declared that every part of Ginger had to be simple, honest, and deep. For instance: "An oval section should be aligned so that its major axis corresponds to the direction of highest loading. A cut line can be beautiful and show distinction between functional elements; a fastener can be a dynamic visual element. . . . No element should have visual appeal as its sole purpose." The guiding principle, cribbed from Honda, was "Man Max, Machine Min"—meaning that the design should maximize room and comfort for the rider while minimizing mass and hardware. "Every cubic inch of 'sailboat fuel' "—air—"should be considered wasted space and an opportunity for further package optimization," wrote Doug and Scott.

Doug had equally strong ideas about how to manage a development team. For years he had been reading and thinking about revolutionary engineering projects and the process of creative collaboration. Now he had a chance to apply his ideas to Ginger. He wrote two philosophical manifestos on the subject: "Ginger Product Development Principles" and "The Ginger Product Development Culture." Every new team member received copies.

Doug often cited an adage from David Kelley, CEO of the design firm IDEO: "Fail fast to succeed sooner." Many managers cringed at this idea, with its implications of defeat and wasted time. Doug considered that view short-sighted and perhaps even dishonorable, because it short-changed both the product and the consumer. He believed that by building and testing many design ideas early in the process, the team would discover their flaws before they became entrenched and caused bigger headaches later. He called these experiments "chicken tests," after the airlines' practice of testing jet engines for failure by throwing chickens into them.

His stint as engineering manager on Fred had taught him a lot about how *not* to run things. "That project has been managed in a way that just blows my mind," he said. "Any other project would have been killed, but Fred can't die because the technology is too good." By that point, the fall of 1999, Fred was more than a year and millions of dollars behind schedule. (It would get much worse. The most recent guess is that the wheelchair, now called the iBOT, will debut sometime in 2003, five years overdue.) The constant setbacks stemmed partly from Fred's complexity as a class 3 medical device, meaning that it involved life and death and therefore was subject to tortuous FDA requirements. But another major cause was the culture Dean had created at DEKA.

He hired highly creative engineers and then encouraged them to pursue innovation as far as their imaginations allowed. "Dean would never do something like give me objectives for the year," said Doug. "He would hate that idea. He would say, 'If I gave you objectives, you might reach them, and that would be terrible, because it might keep you from doing something really great.'"

This philosophy produced innovative devices when five or six engineers worked together on an R&D project that could be completed in a year or so, but it guaranteed chaos when a couple of dozen ingenious engineers dashed off in different directions on a complicated multiyear task such as Fred. For instance, each of Fred's teams had created great components—and then discovered that the parts didn't fit together.

Fred's managers tried to herd their cats, but were often undermined by the wildest cat at DEKA—Dean. "The good is the enemy of the great," Dean liked to say. As one engineer put it, "Dean is sometimes too visionary." Countless times Dean or one of his engineers would come up with a great new idea that yanked Fred in a fresh direction. When a manager vetoed it to stay on schedule, Dean would issue a counterorder, and the happy engineers would scamper off to innovate. Consequently, parts and designs and specs constantly mutated. Too many people had the power to change things, and a change in one component sent shudders throughout the entire machine. It was chaos.

"These are the brightest people I've ever met," said one of Fred's managers. "Everybody here would stand out in any other group. But when you try to unite people, you get slapped down, because so many people don't want limits or restrictions. And Dean likes that. So our strength is our weakness."

All this was especially frustrating for Fred's general manager, Lucas Merrow. Dean regularly undercut Lucas's directives. Lucas had a yellowed *Doonesbury* cartoon on his door in which a thrilled Mike Doonesbury announces to his small software team that he has just found an investor. "All he asks," says Doonesbury, "is that we define a product objective, pick a deadline, and ship on time!" His engineers stare at him glumly, then one says, "Why don't you just chain us to our oars, man?"

"It's really hard to make decisions and drown the rest of the puppies," said Lucas. "You'd think something was decided, and then a few months would go by and somebody who'd been brooding because his little piece had been eliminated would start sneaking it in again, and pretty soon they'd be working on it and I'd have to say, 'Guys! That's *over!*'" He shook his head. "I'm convinced that if it were left to the engineers, with no pressure on them for deadlines, they'd never finish. They would be perfectly happy to keep working on it and improving it."

Fred's beleaguered liaison for manufacturing had a sign tacked to his wall: "There comes a time in every project when you've got to shoot the engineers and start production." Nobody at DEKA wanted to believe that, especially Dean, at least not until Fred had already sucked up $50 million and clearly needed at least $50 million more. Then Dean issued dire warnings about deadlines—but didn't stop making suggestions for improvements. He knew he drove Fred's managers crazy.

"They're afraid when I come down there," he said, "because they're already up to their asses in alligators and all I do is bring new problems." He paused. "New *opportunities*."

Doug had been beaten up by all these forces at Fred, and he was determined not to let them sabotage Ginger. In Doug's view, Fred got into trouble by sticking to the DEKA model, which always asked "*Can it be done?*" instead of shifting to the product development model, which asked "*Should* it be done?"

His plan at Ginger was to encourage creativity at first, then gradually squeeze it off. To explain this to the team, he used the image of a funnel, wide at the top and capable of holding a broad stream of ideas, but tapering toward the point of production. Though a few DEKA engineers had moved to Ginger with him, Doug also made sure to hire people from outside the company who knew how to move from research to product development. He needed engineers willing to drown puppies.

"The list of people who can change things has to keep going down until you freeze the design," he said. "That's the point at which the answer to all new great ideas is 'No!'"

From his time on Fred, Doug knew that "no" wasn't a word Dean accepted well. Doug also knew he would have to deal with that sooner or later, and he wanted engineers who wouldn't make the task any harder.

That was one reason he had hired Ron Reich as the lead mechanical engineer. Ron was capable of drowning entire litters of design puppies. He brought vociferous energy to his task at Ginger. Stocky and dark-haired, with a neat goatee and moustache, Ron came to the project from Detroit, where he had managed sixty engineers for one of the suppliers that orbit the auto industry. Competition among such suppliers was ferocious, but Ron was comfortable bullying his way to the trough and using his elbows to stay there.

That had always been his formula for success: hard work, intelligence, and pugnacity. As a kid he had been proud to be admitted to the elite Bronx High School for Science, an hour and a quarter commute from his home in Queens. He was ranked in the middle of his class, decent but not good enough to get a scholarship. His mother didn't have the money for college, so he lived at home, attended a two-year school, and got a degree in automotive technology. He finished his degree at Fairleigh Dickinson, not known for its engineering program.

He was loud and aggressive, and he slammed into the project at full throttle, running over people in meetings and objecting to almost anything said. On the rare occasions when he did agree with someone, he did so by saying, "I know that." He visibly irritated people, especially Scott Waters, the young industrial designer, whose aesthetic concerns were often bulldozed by Ron's nuts-and-bolts attitude. But Ron did know his stuff. He was in his early forties, which made him an old man on the engineering team and contributed to his know-it-all manner. In the early months of the project it wasn't clear whether Doug would be able to harness him to the rest of the team.

As his lead electronics engineer, Doug had hired another cheerful puppy-drowner named Phil LeMay. Phil had connected to Ginger through Ron Reich; they had worked at the same company in Detroit. A decade younger than Ron, Phil was serious and intense, but also prone to witty quips and bursts of wild laughter. In meetings, he often jiggled a leg up and down as briskly as the needle on a sewing machine.

At lunch one day, the two of them explained why they had left solid jobs in Detroit's car industry for the risk of Ginger. In July 1999 a headhunter had called Ron about a position on a secret project in New Hampshire. The job title—lead mechanical designer—appealed to Ron's dormant love of hands-on engineering.

He visited DEKA. After talking to Doug, Ron returned to Michigan and told his wife, "I just interviewed for a company that doesn't yet exist, that's going to make a product I didn't see—and I can't wait to start." Based on this scant information and Ron's faith, his wife agreed to move their family across the country. A few weeks later, Ron saw Ginger. "I walked around for a few hours with a shit-eating grin on my face," he recalled. He couldn't tell his wife why; Dean insisted on keeping the project secret even from spouses.

Ron quickly felt that his faith had been rewarded. "This is new and it's real engineering," he explained. "*We* determine how fast, how big, how strong. In Detroit the product is so big that it's difficult to see where your piece fits. Here it's almost a parental feeling when you see something at an early stage and get to bring it to development."

Phil LeMay's feelings about moving were more mixed. His wife hadn't wanted to leave their comfortable life in Michigan and still wasn't happy in New Hampshire. "But this is the job I was working toward my whole life," he said. "Before, I had always done *pieces*. I could say, 'See that little box in the car's interior? I did that.' Here I can see the whole thing and say, 'That's mine.' "

Both Ron and Phil mentioned two other lures: the prospect of working with Dean and the expectation of sharing in the wealth if Ginger took off. During their interviews, Doug had mentioned stock options. Now, months later, the subject was still hanging.

In mid-October 1999, Doug called a meeting to discuss packaging. This innocent term was a minefield that could wreck the project. Doug laid out the design issues that he and Ginger's CEO, Timothy R. Adams, had thrashed out. Tim had come to Ginger after decades in the automobile industry, and he knew the hazards of designing and launching a product. Doug knew engineering. Both men had thought a lot about how to manage a team. The engineers were Doug's, so Tim stayed out of the way unless he sensed a misdirection that might complicate manufacturing, marketing, or service, his areas of expertise and focus.

Tim and Doug knew that every one of the packaging issues was difficult and potentially dangerous. To keep the machine small, it needed to be as tightly designed as possible ("no sailboat fuel"). The plan called for two models of Ginger, one for off-road riding and one for cities, but to save manufacturing costs and problems, the models should share components wherever possible. The number of seals and assembly operations should be kept to the minimum. The machine should be easy for licensed technicians to service and repair, but difficult for consumers to tinker with. To save warehouse space and shipping costs, the machine had to be easy to box, and it had to be light enough to pick up and compactable enough to fit into a car trunk.

Today Doug told his engineers that all this was the "outside-in" packaging. It would shape and be shaped by the requirements for the "inside-out" packaging: how to fit the gyros, motors, transmissions, electronics, batteries, charger, and everything else into the chosen design wrapper. Doug drew each of these parts on the whiteboard. He wanted to start with the inner architecture.

Most of the components had to be stuffed into Ginger's chassis, the horizontal platform between the wheels. The engineers called it the "power base" because it was home to the batteries and electronics that powered the machine. The overlord and gatekeeper of the chassis was Ron Reich, the lead mechanical engineer. His rules were severe.

For Ginger to be pedestrian-friendly, its "footprint" on the sidewalk had to be small, about the width of a man's shoulders. That footprint would depend on the size of the chassis. Each component in it had what Doug called "real estate requirements," and each component had territorial advocates among the engineers. In addition, industrial designer Scott Waters would fight to protect aesthetics, and the marketing team would defend the interests of the consumer. The coming months would bring dozens of boundary disputes, squatters' protests, homesteading claims, summit conferences, redrawn border lines, altruistic concessions, and grudging surrenders. Sometimes the result was measured in hundredths of an inch. The team called the main conference space the War Room, which sometimes meant Place of Strategy, sometimes Place of Battle.

Doug began with some basic concerns. "Who plays rock music?" he asked, referring to components that generated electronic noise. He answered himself: "The motors and the transmissions." He drew green lines across those parts on the whiteboard. "And who hates noise?" He drew lines on the gyros and the electronics boards. "Which also *play* rock music," he added. "Those family feuds are your area, Phil."

Phil LeMay grinned at Doug's analogy. Because so much technology had to be packed into such a small package, one of the team's worries was electromagnetic interference, or EMI. Current running through wires creates a magnetic field that can disturb other signals. When you turn on your blender and your radio starts buzzing, that's EMI, which engineers also call noise. EMI can degrade performance or even block out signals, causing failure. In Ginger that could mean loss of dynamic stability, which could mean injury, which could mean lawsuits. One of Phil LeMay's jobs was to keep the noise down.

"What other things are most likely to go wrong?" asked Doug.

"Batteries," said Phil.

"Connectors," said Ron Reich. "They tend to get loose and come off."

"So let's keep them to a minimum," said Doug.

"Electronics," said Phil. "Vibration is a problem."

The first step was to choose a design for the chassis. Doug split the engineers into pairs and gave them a week to come up with design ideas, which they would discuss and vote on. The surviving puppies would be prototyped in hard foam during the following week and voted on again. Doug reminded everyone that in a couple of years, two thousand Gingers a day would be rolling off the factory line, five hundred thousand machines a year. Ginger needed a solid base to build upon. End of meeting.

Over the next two weeks, the small teams sketched possible designs. Twenty emerged. These had been taped to the War Room's walls and examined; then each team member had put a sticky face, either frowning or smiling, onto each sketch. Doug often used visual aids such as varicolored dots, stickies, and pens, as well as charts, graphs, and pictures.

The seven surviving concepts had been roughly modeled in foam. These were now being passed around the War Room and debated in terms of durability, serviceability, potential problems, and other matters. To vote on the designs, each member got three sticky dots—red, blue, and yellow—worth, respectively, three points, two points, and one point.

"Consider cost, assembly, and manufacturing," said Doug.

"And you can't put more than one dot on any concept, J. D.," said Phil LeMay, smiling.

"Is that really true?" asked J. D. Heinzmann, dismayed. He had been lobbying hard for his own design, which slipped the electronics boards into vertical slots. "Then I'm not using my blue and yellow dots," he said, "because I only want one design, number 8."

"Do you know something I don't?" asked Phil.

"Yes, I feel I do," said J. D.

After the dots flew, four concepts remained alive, including J. D.'s. But a few days later, Doug told him his design had come in second. It was too innovative, with too many unknowns, for this stage in the project. Maybe it could be adopted down the road, but at this point it was one of those doomed puppies.

Doug had made the decision with counsel from Ron Reich, the chassis Pig. This classification stemmed from another of Doug's management systems. Doug had once heard an executive say that every project is like a ham-and-egg breakfast in which participants play different roles. The Pig is critical to the breakfast because he has the most at stake, whereas the Chicken contributes but can always walk away. To this confusing barnyard metaphor (were pigs that became ham successful?) Doug added Cows, who were consulted but uninfluential, because they had no stake in the outcome.

For instance, Phil LeMay and J. D. Heinzmann were Pigs on electronics, but were merely contributing Chickens on the chassis, while marketer Mike Ferry was a Cow who had the right to an opinion but nothing more. Ron Reich was the chassis Pig because he was responsible for making it low cost, high quality, and capable of accommodating the other critters' real estate requirements. Depending on which part of the machine was being discussed, everyone got the chance to be a swine, to cluck in protest or support, or to moo without relevance. In this way Doug created hierarchies that were clear yet fluid and didn't chafe. When everyone considered himself a Pig, the result was porcine gridlock. That's what had happened at Fred.

J. D. had accepted Doug's decision about the chassis, despite the conviction that it was a mistake. When J. D. believed in something, nobody was more gung-ho. Dean had recruited him to DEKA through J. D.'s younger brother, Kurt. The tale went that when Dean asked Kurt whether J. D. was any good, Kurt squinched his face and said, "Eehh." J. D. told the story with a chuckle.

In late 1994 he underwent DEKA's notorious interview, in which the candidate stands in front of a whiteboard and gets peppered with questions by the company's brightest engineers. Dean wanted people who not only knew math and physics but could hold up under pressure and be spontaneously creative. He also wanted people who would fit into DEKA's culture of hardworking nonconformists.

Sample question: If a brick heated to 250 degrees Fahrenheit is thrown into an infinite ocean that's 35 degrees, graph what happens to the temperature at the brick's center over time. Some questions were designed to reveal a candidate's thought processes rather elicit an exact answer. For instance: How many pairs of shoes can you get from a cow? Or: If you drink a soda with two hundred calories, how many stairs will it allow you to climb? The interviewers asked questions until the candidate didn't know the answer, to see how he reacted to that condition.

The interview shattered some people. Others panicked. "I've seen a lot of people crash and burn," said one veteran. A few candidates turned belligerent. The worst ones made things up or tried to bullshit their way through or obstinately defended indefensible answers. Such people were crushed without mercy. The best ones kept their cool and tried to solve the problem in unconventional as well as conventional ways. They were self-assured enough to ask for help. Most of Dean's engineers had relished their whiteboard interviews. "It was like devouring a really good meal," recalled Kurt Heinzmann.

J. D. hadn't enjoyed his interview quite that much—he had less of the gladiator in him than Kurt—but he did fine. At one critical juncture the question came down to knowing the impedance of a capacitor. When J. D. said he didn't remember, Dean rolled his eyes. J. D. started thinking out loud. He knew that the impedance increased with the capacitance, so that was $1/c$. And it increased with the frequency, so that was $1/f$. One of the interviewers chimed in that of course you needed a $2(\pi)$ in there to convert radians to cycles. So, concluded J. D., the formula for the impedance of a capacitor was: $1/2(\pi)cf$. Dean smiled. He liked engineers who could think their way through a problem.

J. D. would get plenty of opportunities for that at Fred, where he first worked. Like Kurt, he was wiry and energetic, with penetrating eyes. He was also an inveterate tinkerer. He talked proudly about his grandfather, Horace Raymond, an engineer who invented the Magic Eye door for Stanley Works. J. D.'s father was an electrical engineer. That's what J. D. aspired to be in high school, but when he was about to leave for the Rochester Institute of Technology, his father suggested that mechanical engineering might better suit his skills.

J. D., in his open-minded way, weighed this advice and changed his path. "He was right," said J. D. "I'd spend hours beating my head, trying to fix a broken TV. I loved electronics but didn't understand it. Then in my junior and senior years I took electives in electrical engineering and finally got it, because I saw similarities to mechanical engineering. Voltage was like pressure, current was like flow, resistance was like a shock absorber."

His next enthusiasm was controls engineering. He earned a master's in it at the University of Virginia. While in Virginia, his tendency to plunge into things showed up in other ways. He made 220 jumps as a skydiver, and also became intrigued with Werner Erhard and EST.

After getting his degree, he was invited to work for Erhard's Hunger Project in California. He expected to spend six months there. Six years later he ran out of money from a small inheritance, which forced him to return to engineering. "It was the best thing that ever happened," he said, "because I had to depend on myself."

He worked on a dental robot, then joined an engineering services company. But he missed doing real projects, so when his brother Kurt called, he was ready to jump. He started at DEKA in June 1995. His wife stayed in Marin County. She loved it there, and J. D. understood her reasoning. They saw each other every few weeks. He liked the arrangement because it gave him interludes of intense romance followed by expanses of uninterrupted work. Perfect for DEKA. He was also back with his first love, electronics.

When Dean moved him to Ginger, J. D. was thrilled. The very word Fred caused him to bury his head in his hands. He was revved up about working with Phil LeMay on the electronic architecture for Ginger. Soon he might take on the boards that drove the motors. "We need to integrate the technology for driving the motors and for balancing the machine into one module," he said. J. D. didn't know how to do that, which made the job doubly attractive.

Over his desk, photos of his scientific heroes formed a Wall of Electrical Fame: Watt, Volta, Gauss, Coulomb, Maxwell, Henry, Joule, Ampere, Faraday, Morse. He loved to read the biographies of these trailblazers, and histories of engineering. He also had definite views about the history of Ginger, in which his brother figured prominently.

"Kurt has made a huge contribution in terms of speed, range, hill climbing, motor size, and battery size," he said. "As far as I'm concerned, he invented Ginger."

Kurt certainly helped to dream up Ginger, but J. D. and the rest of Doug's team had the tougher job—meeting the dream's deadlines.

chapter three

CEO Mode

October–November 1999

Through FIRST, Dean had met François Castaing, the legendary director of engineering at Chrysler. Castaing was part of the small group that in the 1990s transformed Chrysler from a sluggish giant into a cocky, nimble innovator, without question the most inventive car company in the United States. Castaing championed change and radical ideas, so naturally he and Dean hit it off. The Chrysler executive became FIRST's greatest advocate and fund-raiser among the industries in Detroit. He soon joined FIRST's board of directors and continued to work hard for the group even after retiring from Chrysler in January 1998.

So in the fall of 1998, when Dean was forming plans to develop Ginger, he asked Castaing to recommend someone with experience in big-time manufacturing. Castaing suggested his former colleague, Tim Adams. Tim was president of Chrysler Europe, but like many of the car-maker's top executives, he had been restless since Chrysler had merged with Daimler-Benz earlier that year. In October 1998, Castaing set up a meeting between Tim and Dean at the Rattlesnake Club in Detroit.

They sized each other up and liked what they saw. Tim had the background Dean was looking for as well as extensive contacts among suppliers in the United States and abroad. Dean intrigued Tim. So did the prospect of building an international transportation business from scratch. But Tim told Dean that one thing made him leery. He had once tried working for a mercurial entrepreneur and hadn't liked it, because the owner had insisted on retaining control over every little detail and would sell the desk out from under you for cash flow. Working for him had been chaotic and suffocating, Tim said, so he wanted to be sure that Dean really did intend to let go of the reins. Dean was adamant—he had no interest in being involved with manufacturing and all those production details. He needed someone like Tim to run the business so he could get back to the fun stuff, inventing and engineering.

Tim believed him. Dean probably believed himself. Tim visited DEKA and saw Ginger. That sealed it, though the negotiations over his employment package dragged on. "Frankly, Dean and Bob Tuttle wore me down," said Tim later. Tuttle was Dean's financial guru and most trusted advisor. "They're *very* good," said Tim with a grin, "and Dean's a charming guy." Tim joined Ginger as CEO and president in early 1999, under terms that dropped his income more than 50 percent. As a compensatory perk, he asked to fly business class whenever he traveled, which Bob Tuttle approved.

Tim spoke in short, decisive sentences, and had the hurried air and manner of a no-nonsense executive. He always walked fast, head down and eyes focused, as if he knew exactly where he was going. He looked like an executive as well. His wardrobe, though casual and understated like everyone's at DEKA, stood out for its luxe stylishness. He kept his moustache and thinning hair meticulously trimmed. He also had decisive executive ideas about how to do things and launch the business, such as hiring experienced senior managers and offering them stock packages.

Much of this was new at DEKA, and irritating to Dean. That first summer, as Tim started trying to hire top people and give them options, Dean balked, sending word through Bob Tuttle that Tim needed to rethink his offers. That gave Tim pause. So did one of Bob Tuttle's financial forecasts for Ginger, which assumed profitability in the new company's first half-year.

Yet Dean kept amazing him with his charisma and his endless Rolodex of contacts. The excitement of the start-up energized Tim, too, and Doug Field seemed to be putting together a great team of engineers. By the time the leaves began turning color in the fall of 1999, Tim knew that the honeymoon with Dean was over. But he could be patient and learn to adapt.

At 10:00 one October night, Dean and I went to a scruffy blond-wood joint in Bedford called T-Bones. "The only place I know that serves food late," he said. Dinner with Dean was always late. He was often the last person to leave DEKA, and he would walk through the building turning out lights like a parent. He hadn't eaten or slept much for two days and was starving. He never cooked for himself, and over the course of a year and a half, I saw him eat only one meal at home with his girlfriend, K. C.

The waitress at T-Bones greeted him by name and asked the cook to relight the oven to melt the cheese for Dean's French onion soup. Talking over loud, piped-in rock and roll, Dean said he had just returned from speaking at Agenda 2000 in Scottsdale, Arizona, a gathering of more than five hundred top executives, mostly from dot-com companies. He had wheedled nearly $200,000 out of them for FIRST. They also had given him a standing ovation for Fred, but they couldn't understand why getting the wheelchair to market would take two more years, an eternity in dot-com land.

"I've been working on it for longer than most of their companies have existed," Dean said. "The Internet is like sports. It's fun and expensive and it *doesn't matter.* If Amazon or Infoseek disappeared tomorrow, no one but a few employees and stockholders would care. As opposed to things that do matter, like air, water, food, shelter, health." To him the so-called New Economy looked like a house of cards built on a bubble. He didn't understand why creative people, especially engineers, devoted themselves to bubbles rather than to real products that improved the world. When the bubble burst, as Dean suspected it would, all those dot-com stock options would evaporate. Real products, the kind that had once made America great, didn't disappear.

Uncharacteristically subdued, he was tired and had problems. Morale at Fred had been sinking. The project had been delayed, then delayed again, then again and again. After putting $50 million into the wheelchair, with no end in sight, Johnson & Johnson was getting antsy and intrusive. Fred's chief systems engineer had recently quit in frustration and taken a job at a dot-com company that offered generous stock options. "Not a good sign," said Dean over his soup.

Resentment of Ginger also had been growing at DEKA. Most people at the company had never been through the locked door with the circular black and white sign that said "Project Ginger: DEKA Authorized Personnel Only." Though Ginger was adopting technology from Fred, even most Fred people weren't allowed into the new project. That grated. So did the fact that Dean planned to turn Ginger into a semi-independent company whose employees eventually would share the wealth. No other group at DEKA had ever gotten such an offer.

Ginger was breaking DEKA precedent in other ways, too. The project didn't have a dime of its own—Dean was still funding it—but some DEKA old-timers felt that the Ginger crowd already considered themselves a bit more important, the chosen few. Ginger had an industrial designer, a prettification that no one at DEKA, including Dean, had ever considered necessary. Ginger had a CEO, for chrissakes, a bigshot lured away from Chrysler—Chrysler!—and the CEO had hired a couple of marketing guys.

CEOs, industrial designers, and *marketing guys*? On Dean's payroll? No wonder he wanted to use somebody else's money to fund Ginger. It was hilarious in a way, but also irksome to veteran DEKA engineers. What would they do next over there? Draw an organizational chart? Wear neckties and go home at five? At DEKA, people liked to say, the organizational chart consisted of Dean and the company phone directory. Ginger was changing that. Dean worried about these resentments and shared many of them. He encouraged independence—but not from *him*.

On this October night at T-Bones, he was more concerned about money. Ginger was burning cash like dry leaves. For months, he had been courting investors, showing rough prototypes of the machine to a parade of millionaires, billionaires, and venture capitalists. He was looking for $50 million in funding, for which he was willing to part with 10 percent of Ginger. He couldn't find anyone willing to put up that kind of money on those terms. One VC had offended him by demanding 20 percent of Ginger up front, before raising a nickel. But that's the way things were being done at the moment, when VCs expected big bites of up to 50 percent of anything they funded, usually some sort of soufflé with the word Internet attached to it. It affronted Dean to be grouped with such lightweights.

I never tired of watching Dean perform his sales pitch to potential investors. It was entertaining and irresistible. Engineers know that heat always flows from a hot object to a cold object. When Dean began talking about his passion for Ginger to a cool investor, you could watch the target's molecules heat up and start dancing.

Dean believed wholeheartedly in doing well by doing good, so he opened his pitch with an aperitif of statistics that sharpened his listener's appetite for more: Do you realize that transportation is the world's biggest industry, bar none? And that transportation and pollution are two of the world's biggest problems? Sixty-five percent of all airborne pollution comes from automobiles, yet only 6 percent of the world's population owns cars. So what will happen when billions of Chinese and Indians can afford automobiles? Disaster! Unless we have an alternative. Even if we *did* have feasible, affordable alternative-energy cars, which we don't, they would solve only part of the problem, the other part being traffic jams. Our cities can't handle any more big vehicles. Yet in fifteen or twenty years, most of the world's population will be living in cities. Did you know that China is building eight new cities the size of Manhattan every year? Cities are *dying* for solutions.

After ten or fifteen minutes of this, when the target began to realize that he'd heard this speech before, just not so entertainingly, Dean would tell the story of Fred, beginning with the wheelchair-bound guy in the mall. Then he would take the potential investor into his projection room and run a segment that ABC's *Dateline* had done on the machine in June.

Ah, a product, something an investor could get his teeth into. That one's already taken by Johnson & Johnson, Dean would explain, but DEKA owns the technology and its nonmedical applications, which go far beyond Fred. More bread crumbs leading to the snare. Even if our cities weren't dirty and crowded, he would say, even if the billions of Chinese and Indians never could afford mechanical transportation, there are still huge markets for the product you are about to see. Did you know there are 1.1 million mail carriers in the United States? A couple million delivery people? 13.6 million kids on college campuses? Did you know that golf carts alone are a $4 *billion* a year business?

Then he would put in the Ginger video, shot by a professional acquaintance one Sunday at Dean's house. It began with images of old modes of transportation and led to photos of traffic jams, as the Talking Heads sang "Road to Nowhere" in the background. Then came a voiceover of Dean lecturing about everything the investor had already heard, except this time Dean was cruising along on Ginger. There were vignettes of Doug Field, Benge Ambrogi, and other engineers going about their business on Gingers, dressed as soldiers, guards, golfers, or deliverymen. They bounced down stairs, raced up a hill, and bumped over rough terrain while the Beach Boys sang "I Get Around" and the B-52s sang "Roam." An older woman rolled along, too—"my mother," Dean would explain. "She's seventy-five and we put her on it."

When the lights came up, the investor would look very interested. Dean would say that he could have gone to a golf cart company and made a lot of money, but he wanted to change the world. "We want to sell in a lot of countries, on a lot of continents," he would say. "We want to be the fastest-growing company in the world."

Then he and the investor would walk down to Dean's double helicopter hangar, where Dean would tear around on a Ginger until the target's eyes got big. Then Dean would switch the machine into a slower mode—known among the engineers as "CEO Mode"—and give the prospective investor a quick lesson. Dean liked to demonstrate Ginger's balance by shoving the billionaire rider in the chest: "See? *See*?" he would say, jabbing the guy hard. "It automatically compensates." Then the rider would roll off and his face would get that Ginger look that meant the hook had gone in deep.

But once they heard Dean's terms, they snorted in disbelief. Dean shrugged and bid them good-bye. Eventually Dean turned to one of his contacts from FIRST, Paul Allaire. Allaire was the former CEO of Xerox and now served as chairman of FIRST. When Allaire saw Ginger, he wanted in, and offered to find investors among his friends. These included N. J. "Nick" Nicholas, former president and CEO of Time Warner Inc., who got so excited by Ginger that he arranged for his brother Pete to visit Dean two days later. Peter Nicholas was cofounder, chairman, and CEO of Boston Scientific. In 1999, when he visited

Dean, *Forbes* listed his worth at $2.4 billion. Dean gave him the spiel and jabbed him in the chest like all the others.

It dazzled Dean that billionaires were coming to *him,* and confirmed his opinion that Ginger was world-shaking. By early fall of 1999, five individuals, led by Paul Allaire, had offered to give Dean $5 million each in return for one percent of Ginger. So had two or three of Dean's other business friends. The investment/percentage ratio was based on a pie-in-the-sky valuation of Ginger, guesstimated by Dean and Bob Tuttle, at half a billion dollars.

As we sat in T-Bones, Dean said that the deal with Allaire and the others was supposed to close in a few days, but Xerox's stock had dropped nearly 40 percent in the past week, and Dean suspected that Allaire's options were almost worthless. If Allaire backed out, the others might follow him.

Dean never put all his eggs into one basket, though, whether with a supplier or a client. He also had been talking to John M. "Jack" Hennessy, former chairman of Credit Suisse First Boston, one of the world's largest investment banks. Hennessy had retired in 1996 to become director of CSFB's private equity division, overseeing a fund of about $3 billion for investment in emerging companies and technologies. Hennessy loved Ginger but wanted the usual absurd percentage. Such things were like flirtations. Usually the small new company played the pleading suitor, while the investor took the role of the distant trophy.

But Dean wouldn't use that script, because it allowed investors to end up owning a controlling percentage of the company. He reminded Hennessy and every other potential investor that DEKA wasn't an unproven start-up, nor was Ginger some nebulous dot-com concept. It was already a real product with five years of R&D behind it. In Dean's view the biggest question—can it be done?—had been answered. That meant lower risks for investors, and *that* meant a lower percentage of ownership.

Hennessy kept calling to tell Dean that his terms were crazy—and to ensure that Dean wasn't making a deal with anyone else. Dean kept Hennessy on edge with hints about the enthusiasm of other prospective investors. In general, the longer a start-up goes without funding while it burns through cash, the more desperate and pliable it becomes when seeking investors. At T-Bones, Dean speculated that Hennessy was counting on that. It was a poker game. Hennessy didn't understand him yet.

A week later, Dean was down at Fred talking to Benge Ambrogi. They were going to Dusseldorf, Germany, the following week to show off the wheelchair at the world's biggest trade show for rehabilitation equipment and products. Dean suggested that Benge might want to stop at British Aerospace Engineering (now BAE Systems) on his way back to talk about gyros, which both Fred and Ginger used. Benge said he didn't need to, because John Morrell was flying to BAE in a few days. John had been Fred's lead controls engineer before moving over to Ginger. Dean wondered why John needed to go, since Doug Field and Tim Adams from Ginger were already at BAE this week.

"I don't know," said Benge. "I just know John is going." Then he added, with raised eyebrows, "And it's a $2,500 ticket."

"$2,500?!" said Dean.

"That's Ginger," said Benge, shrugging. His own ticket, he added, had cost $1,100. "But I'm Fred, not Ginger," he said. "Maybe that's a bad attitude, but that's the truth."

Someone had told Dean that the Ginger guys, from CEO Tim Adams on down, had been flying business class everywhere (a false rumor), and that Tim even flew business class from Boston to New York. As he walked away from Benge, Dean was bristling. "I hired this big guy with all this experience," he said, referring to Tim Adams, "and he's going to be running this big company, and I've been trying not to sit on him. But we're going to fix this." His fuse burned a little farther. "I have *never ever* flown in the front of the plane, because it doesn't make any sense to me that a seat up there costs $1,000 more than a seat a few rows behind it. It's the airlines' way of charging idiots."

Most of the time, of course, Dean did sit in the front of the plane—while piloting his private jet. But it didn't seem like the right time to point that out. Few things made Dean fume like people spending his money on things he considered extravagant.

The subject was still bothering him at dinner that night in a small rib joint near DEKA. "That $2,500 ticket, which I heard about four times today," he said. No wonder people at DEKA resented Ginger. "It is *my money* that Tim is spending," said Dean. "He doesn't seem to understand how a start-up company works—you don't have lots of money to spend, because you don't have a product. Sure, I've got a lot of toys," he added, meaning his two helicopters, his Citation jet, his Humvee, his Porsche, his private island, and his fabulous estate, with its sauna, wine cellar, indoor pool grotto, projection room, regulation baseball diamond, basketball court, and antique elevator from the movie *The Sting*. "But I've always waited to see whether the product was going to make money. If it did and I paid everybody off and sunk more back into the business and I still had money, *then* I would buy toys and splurge."

Work hard, watch your pennies, turn out lights, earn your splurge. "I have to talk to Tim," he said grimly.

His mood wasn't improved by the slide of Xerox stock for the second straight week. He still hadn't called Paul Allaire about the Ginger investment, afraid that some of the other investors might not stick if Allaire wasn't involved. He had started working on Plan B: a new set of investors.

When Johnson & Johnson decided to reveal Fred to the public, the wheelchair needed a grown-up name. Dean christened it the iBOT, but everyone at DEKA, including Dean, still called it Fred. The machine had emerged from secrecy in late June, in a segment by John Hockenberry for *Dateline*. When Jane Pauley referred viewers to the show's Web site for further information, it crashed within minutes, overwhelmed. The story registered thousands of hits within twenty-four hours and was so popular that *Dateline* took the unusual step of repeating it in August. (When his paranoia was running higher than usual, Dean worried that Japanese engineers were examining the tape frame by frame, and were bound to make the leap to Ginger.)

But Fred had never been out in public until the trade show in Dusseldorf. Every twenty minutes, two DEKA engineers put the machine through its repertoire at the Johnson & Johnson booth. Word spread quickly, and soon the crowds were so deep that several neighboring exhibitors demanded to be moved. Some people sat through three or four shows. Ovations were standard. The invention starred on the local evening news.

"Fred wasn't the hit of the show," said Dean. "It *was* the show. I was really *really* bad," he added, laughing. He drove around the exhibition halls on Fred, wreaking havoc and gathering followers. At booths where conventional wheelchairs were laboriously climbing into special vans or inching along uneven terrain, Dean zipped onto the scene, nimbly executed whatever task was being tediously demonstrated, then saluted his stunned competitor and rolled off.

"It was like being in a time warp," he said, "like walking around and seeing Victrolas next to CD players." He laughed again. "I've been waiting so long for this. I really needed it. And that team needed it. If I had known what sort of reception it was going to get, I would have shut down Fred and taken the whole floor with me." He thought he might do exactly that for the next big trade show in June, in Paris. "I'm going to do the Champs on Fred," he said, pronouncing it Shawmps, "from the Arc de Triomphe to the Louvre." The idea made him grin.

"You talk to him and he says a few bright things, but he says the same few things every time," said Dean one night in November. The subject was Tim Adams, Ginger's CEO and president. "He's a simple guy in a lot of ways, and that's his strength." Dean sounded like he was trying to convince himself. "He's not nearly as bright as most of the engineers I have, even though I pay him two or three times as much as any of them."

Then what was Tim's value to Ginger? "When it comes to knowing suppliers and manufacturing people," said Dean, "he's been there, done that, and knows everybody."

Dean's assessment wasn't fair to Tim, but was probably inevitable given their different work histories. Both were stars in their particular orbits, but they had been orbiting different planets. DEKA was small, provincial, often cash-poor, unbureaucratic, dedicated to innovation, and dominated by one visionary engineer/entrepreneur. Almost immediately after Tim started at Ginger, Dean pigeon-holed him as a typical product of the U.S. auto industry, which he perceived as spendthrift, bureaucratic, dedicated to incremental changes, and dominated by M.B.A.'s and stockholders.

That may have been true of most American car companies, but in the previous decade Chrysler had been an exception in the domestic auto industry, and so was Tim. Unlike Dean, Tim was a cosmopolitan who had worked on three continents, always in the auto industry. At fifty-three, he was five years older than Dean. Early in his career he was managing director of Jeep in Venezuela, then director of South American operations for American Motors. He moved to Chrysler and became president of a joint venture between Chrysler and Renault, then president and CEO of Alfa Romeo Distributors of North America. When Chrysler asked him to oversee the sale of its disastrous Lamborghini subsidiary, he instead made it profitable within a year. Next he became director of Chrysler's New Generation Vehicle Programs, then president of Chrysler Europe, during which he made six acquisitions in eighteen months. He also had negotiated for Chrysler in China.

Not the credentials of a Rust Belt dullard. Tim had managed two start-ups and three turnarounds, and was accustomed to commanding hundreds of employees, making quick decisions, and not worrying too much about nickels and dimes. But he soon learned that at Ginger every important decision, and many unimportant ones, had to be cleared by Dean. It was exasperating. Yet Dean and the DEKA system were so intriguing, and the potential of Ginger was so great, that Tim was still willing to be patient.

He felt bad about the brouhaha over the business-class tickets. Despite Dean's assumption that Tim had picked up sumptuary tastes in the auto industry, he had only flown business class at Chrysler on long international trips. When he joined Ginger he had asked for the perk to help soften the big salary cut, and he thought Dean had approved it. He had no idea Dean felt so strongly in the negative, and he didn't mind giving up the perk. In hindsight he realized that flying business class had sent the wrong signal in a start-up company. Tim was willing to learn and to correct his mistakes. He assumed that Dean was, too, and that eventually he would loosen the reins.

"Tim has good ideas and wants to run it," said Lucas Merrow, Fred's general manager, who spoke from experience. "But I don't know if Dean will let it go."

chapter four

Frogs

November–December 1999

S cott Waters, the industrial designer, and Kevin Webber, a design engineer, were talking about design, color, and texture for the area on Ginger where the vertical transmission covers merged into the horizontal chassis, or power base. The mention of the chassis drew the attention of Ron Reich, one desk over. "It would be better if it were shot-blasted," said Ron, honking his way into the discussion and drawing on his experience in Detroit. "A nice shiny finish will be beautiful for the first one, but less beautiful at 250. By 50,000 you'll be getting heat-checking marks and you'll need a lot of polishing and handwork."

Scott barely looked over at Ron. To Kevin he said he was looking for ways to get color into the machine.

"If you use paint," said Ron, continuing to honk, "it may look different depending on the substrate, whether it's plastic or metal."

"I'd like to avoid paint," said Scott, "but I would like color on the wheel. I'm afraid it won't look like quality if it's just a plastic wheel."

"What about a cap on the hub?" asked Ron, showing his Detroit roots again.

"*No,*" said Scott. "Every part of this machine needs a function, and that wouldn't have one." He pointed to the flanged curve between the wheel and the chassis. "This piece is not just cosmetic. It protects the user's legs from the wheel."

Kevin wondered whether they still expected to attach the wheel with one center bolt or needed three. "I'm thinking about assembly issues," he said. "Those piddly things that happen after Scott designs it."

"See those gray hairs on Kevin's head?" said Ron. "Half of them are from assembly issues."

"And the other half are from design issues," said Kevin.

Scott was still thinking about color. "The Michelin guy told me they have a hundred different blacks to choose from," he said.

Kevin stared at him. "On the day you discuss tire color," he said, "I'm going to be out."

Kevin Webber was a consulting design engineer who had landed a two-week gig at Ginger in the spring of 1999. He proved too valuable to let go, and six months later was still at the project full time, with no dismissal in sight. He was the conduit between industrial design and engineering, meaning that he functioned at various times as mediator, shield, and referee. He played these roles with a bottomless supply of patience and humorous quips. A cartoonist might have drawn him holding off a charging Ron Reich with one arm and a windmilling Scott Waters with the other, while cracking jokes about both. A more realistic image would have shown him staring at a CAD drawing on his computer, Scott at his shoulder, searching for a way to bend the principles of engineering to embrace one of Scott's designs in a way that Ron wouldn't crush.

Scott Waters, like many of DEKA's engineers, had a professional pedigree. His father was an industrial designer; his grandfather had been a design engineer. He had grown up in Nashville, but his accent had been polished off by years of college and work in the North. Before Ginger, he had designed for a small firm in upstate New Hampshire that had done some work on Fred. He had designed furniture, snowboard bindings, and a toilet for American Standard. He was twenty-eight years old.

"Before I took this job, I was warned about working with engineers," he said. "Other designers told me, 'They just suck the creativity out of you,' and 'You'll have to fight for every decision.' But Doug is committed to design and has an amazing eye. And we share this . . . this *vision* for Ginger. Doug inspires me. I never thought I'd say that about an engineer."

One of Scott's heroes was Jonathan Ive, the designer of the iMac. Ive understood that technology doesn't come with a prescribed form, that a designer has freedom as well as responsibility. Until the iMac, Scott explained, a computer was just a sheet-metal box with slots, whose purpose was performance. Ive made the technology visible, lovable, *fun,* and by doing so changed the technology's meaning. People responded by buying two million iMacs in one year. "Apple puts a lot of passion into things," said Scott. "Their products take chances and have a lot of heart. I want that for Ginger."

Ive had the advantage of reacting to antecedents. Scott and the rest of the Ginger team were creating history from scratch. Scary, but that offered an advantage of its own: Scott didn't have to use bells and whistles to differentiate his product from similar ones, because there were no similar ones. "We want it to be simple, unique, and high performance," said Scott. "A tool for mobility. Doug says it's like putting wheels on your feet."

For inspiration Scott looked at his "mood boards," large laminated collages of products whose design he admired: phones, bikes, sunglasses, cameras, stylized motorcycles. Most cars, in his judgment, were "blobjects" with no definition, no drawable profile. Compare that, he said, to the Audi TT, whose form was seductive yet hard-edged. He hoped the same would be said about Ginger. "It boils down to longevity," he said. "You want a design that lasts."

He also wanted something airy, sleek, and easy to read, like the Aeron chair. In an earlier version of Ginger, he and the engineers had blundered by enclosing the wheels, motors, and transmissions in big housings. "We created a huge wall around the platform, which gave it a boxy feel, very confining." The rider felt squeezed into a chariot. "To maximize the space, we shrunk that package *into* the wheel," said Scott, excited. "Now the wheel starts to *define* the machine. It creates depth and interest, and it connects the platform with the wheel assembly."

The handlebar worried him. So many things did. Ginger would be a defining moment for him. Could he fulfill Doug's vision? How would the design be judged?

"To me this is the opportunity of a lifetime," he said earnestly. "It's not just designing a clamshell for something electronic. It has all the meat I could ever want: concerns about structure, cost, worldwide acceptance, and usability."

So far Dean hadn't said much about his design ideas. The boss's favorites seemed to be Scott's quick sketches of police officers on Gingers equipped with windscreens and antennas. Dean always showed those to the potential investors who visited.

In early December, Doug's whiteboard said "69 Weeks Until Launch." It was also Frog Day. In typical fashion, Doug had taken one of Dean's casual aphorisms—"You gotta kiss a lot of frogs to find a prince"—and methodized it into a monthly management tool. On Frog Day you couldn't work on anything you knew. Today, for instance, the electrical engineers were designing a kickstand and the mechanical engineers were investigating wiring. Each team had to make something that addressed one of Ginger's problems in an original way. Failure was encouraged because it implied real daring and creativity. On Frog Day, Doug wanted full-lip smackers, not safe pecks.

"The most spectacular failure gets the Frog Award," said Doug, "for pushing the envelope the farthest." The award, a large cast-iron amphibian, would soon be squatting on some engineer's desk.

The products of Frog Day revealed the fine line, and often the not-so-fine line, separating ingenuity from preposterousness. Frog Day generally produced plagues of ugly bug-eaters, such as a steering mechanism made from a tennis ball and the kickstand that electrical engineer Phil LeMay was working on at the moment. Phil had cut some levers from plywood, angled them up from the chassis, and connected them with a red string that ran over the handlebar to a wooden dowel in a steel tube screwed on with brackets. That description sounds confusing, but it's accurate. Phil's froggy kickstand was what engineers call a kluge, once defined as "an ill-assorted collection of poorly matching parts, forming a distressing whole."

Phil was being helped, theoretically, by a marketer named Matt Samson.

"So we're bound to succeed," said Phil to Ron Reich.

"No, you're bound to fail," said Ron, expressing the standard engineering opinion of marketing.

"But we'll fail in a way that the consumer wants," said Matt.

According to Doug, one of two things happened on Frog Day. Say an electrical engineer was frustrated because the mechanical engineers weren't solving some issue related to his work. The frustrated person got the therapy of trying out his own idea, or learned firsthand why the other engineers were having a problem.

There would be fewer Frog Days as the funnel narrowed and Ginger got closer to production. "Invention is a process of creation," said Doug, "but development is a process of destruction, of narrowing down and compromising. It's like the two aspects of Darwinism: Evolution occurs through survival of the fittest, but also through mutations. DEKA is founded on mutations, not survival of the fittest, but that's what you need to bring a product to market."

Doug looked thoughtful. "Dean provided the spark," he said. "I can't do that. And I can't compete with Benge and Kurt at making mutations. I don't know why." He sounded wistful. "But I *am* good at leading engineers to develop a product." He moved off, in search of princely frogs.

That afternoon Tim Adams called a "town hall meeting" with the entire Ginger team, about eighteen people. Tim liked to keep everyone informed about the road ahead, including potential potholes and hairpin turns, because he believed that secrecy and uncertainty hurt morale. He told the group about his ideas for the company's culture: accountability without much structure or bureaucracy, open communication, teamwork, and speedy decisions made down inside the organization, where people knew the answers.

For manufacturing, Tim said, the company would use the "demand-pull" model, meaning it would build to fill orders, with little inventory.

The first factory would be built in New Hampshire but designed so that "cookie-cutter facilities" could be replicated in Asia, Europe, and elsewhere. Launch was scheduled for early in the second quarter of 2001. In the first ten years, Tim and Dean expected to build about 22 million machines. For pricing, the goal was a few thousand dollars for the all-terrain Pro model and less for the smaller sidewalk-friendly Eco.

One hypothesis proposed by Dean's financial guru Bob Tuttle had the business becoming profitable in the first half-year of production. "That would make us unique," said Tim, "and I'm skeptical." The same plan called for a price/earnings ratio of 50:1, with market capitalization of $26 billion by year five and $73 billion by year ten. Those figures might be optimistic, too, but one thing wasn't in doubt: "This business has the potential to be *very* big," said Tim.

By the launch date they would be burning a million dollars a month. At that rate, said Tim, they couldn't afford delays, which would also hurt their market capitalization. Delays also increased the risk that other companies would hear about Ginger. "People will be trying to attack us," said Tim, "and people will be trying to catch us." Dean and the recent flap over airline tickets hovered overhead as Tim urged "resource restraint" and sensitivity to "cash conservation."

He introduced the new director of manufacturing operations, Don Manvel. Balding and soft-spoken, Don always talked with a half-smile and lowered eyelids. He had spent more than twenty-five years at Chrysler and resembled his friend Tim in other ways as well: He had worked all over the world—Italy, Austria, Spain, Germany; he had taken one look at Ginger and decided to trade security for excitement; and he stood out by being middle-aged.

"It's a brand-new environment for me," said Don, "very refreshing." The first plant would make two thousand machines per day in two shifts, he said. He liked to keep the manufacturing process as simple as possible. "Manufacturing wants any color as long it's black," he said. "Marketing wants two hundred different colors. The engineers want something reasonable."

Mike Ferry, director of marketing, spoke next. He was six-foot-five and a bit gawky, with blond hair that fell boyishly over his forehead. He had joined Ginger two months ago from Procter & Gamble, where he had run marketing for a paper-products division. Because marketers are social creatures who need to cluster in colonies, he had quickly hired two former colleagues, Matt Samson and Tobe Cohen.

In the manner of marketers, Mike invited everyone to share their thoughts with him. Then he began a long presentation. Ginger was a huge marketing challenge because, unlike diapers and paper towels, it didn't have built-in customers. A preliminary marketing study by Arthur D. Little had estimated that Ginger would sell 31 million units in the first ten to fifteen years, a large portion of it outside the United States.

"Europe and Asia are architected much better for Ginger," said Mike. Countries in such places had dense city centers with little parking, and were more accommodating to alternative vehicles, with bike lanes and car-free zones. Taiwan alone had 8.5 million scooters. Getting Ginger on the sidewalks in the United States would be a harder job.

The marketing team had been doing some theoretical frog-kissing of its own and had identified three broad categories for Ginger use: (1) pure recreation; (2) a fun way to get from A to B (target audiences would include college students, hip commuters, golfers, tourists, and shoppers); and (3) an efficient way to get from A to B (target audiences would include commercial customers such as delivery services, the police and the military, people in airports, and workers in gargantuan warehouses or factories).

Which category should they target first? Mike was leaning toward "a fun way to get from A to B," because practical consumers needed reasons to spend a few thousand dollars, whereas fun-seekers didn't care about cost. He outlined other traits within the target demographic: urban, male, aged eighteen to thirty-four, independent, self-confident, trend-setting. Mike explained that his team hadn't done anything they considered real research, such as focus groups, because Ginger was in stealth mode. But he suspected that the marketing plan wouldn't rely on traditional ads, because what excited people about Ginger was riding it. The key to sales would be to get people on the machine.

Ron Reich asked how Mike planned to do market testing without breaking the project's confidentiality. Everyone knew that Dean wanted to keep Ginger secret for as long as possible, and then a little longer. For Mike it was a frustrating, delicate subject, and he hesitated to answer Ron because it might imply a criticism of Dean. Tim stepped in. "We're going to have *no choice* but to do these tests," he said. Mike hoped to start them in the summer of 2000, but no one could predict when, or whether, Dean would give the OK.

One marketing opportunity did interest Dean. Right after the town hall meeting, Tim Adams, Doug Field, Mike Ferry, and several engineers gathered in the conference room near Dean's office. The room was lined with framed photographs of previous DEKA inventions and was dominated by a large oil painting of Dean, done by his father, in which the company founder posed in jeans and a denim shirt, with a full beard (since shaved). After Dean entered and sat beneath his portrait, Mike Ferry started a video. Tom Cruise popped onto the screen, using an Apple Powerbook in *Mission Impossible.* Tommy Lee Jones followed, wearing Ray-Bans in *Men in Black.* Next came Pierce Brosnan as James Bond, getting a demonstration of a specially equipped BMW Z3. Then a boy broke the ice with E.T. by offering Reese's Pieces.

The video ended. All the engineers except Doug looked baffled. Dean began to explain. He didn't recognize any of the actors, but he understood the concept of product placement. Not long after *Dateline* aired the story about Fred, someone had called DEKA from Steven Spielberg's office. Spielberg was about to start a movie called *Minority Report,* starring Tom Cruise and set in the year 2084. One of the main characters used a wheelchair, and Spielberg wanted the iBOT.

Dean had had other business in California, so he had met with Spielberg's people in a room lined with the movie's storyboards, including a sketch of a futuristic wheelchair. "It looked like fins on a baby stroller," said Dean. "A cross between art deco and horseshit." Spielberg's people used the storyboards to summarize the plot for Dean and to exhibit its futuristic visuals, including some clunky capsules that were supposed to be high-tech people-movers.

They asked Dean what he thought of their portrayal of the future. You don't want to know, he said. But they did, they really really did. So, Dean now told his Ginger team, this is how the conversation went:

"It looks like something a ten-year-old could imagine. In short, it's pathetic."

"Do you have something better?"

"Yes."

"Can we see it?"

"No."

"Is it like an iBOT?"

"Is a Model T like a Ferrari? I won't tell *you,* but I will tell Mr. Spielberg—under a confidentiality agreement."

Dean told the movie people that if they used DEKA's secret technology—Ginger—it would be the coolest thing in the movie and wouldn't cost a million bucks for animation.

A few weeks later, all the storyboard people were fired and new ones started over. Spielberg's assistants kept wheedling Dean about the secret technology. The prospect of being at the center of a Spielberg movie made Dean's chops water. "For all the money in the world," he told the Ginger team, "we couldn't buy ads to do what this would do." But two things bothered him. He was afraid that Spielberg would let his minions handle it, which could compromise secrecy, and that Spielberg might expect DEKA to pay *him.* "And I have a different perspective," said Dean. "Namely that Mr. Spielberg should pay *us.*"

Spielberg's people had told Dean that he would hear in a couple of weeks about meeting with Spielberg himself, probably by the end of January 2000. Dean felt sure that Spielberg would love Ginger and want machines by the end of February, probably between three and twenty of them. That would require the team to react very quickly.

The room fell silent. If they kissed this frog, it might turn into King Tut or King Kong, so the engineers didn't know whether to whoop or weep. "We need to make an assessment of what resources it would take," said Tim soberly, "because Doug and I both feel it's a threat to our launch date. If his movie launches in April 2001 and hits 30 million people by June, but we're not in the marketplace until the fall, we've blown the biggest chance we'll ever have."

Dean disagreed. Johnson & Johnson had told him it was impossible to buy the kind of exposure *Dateline* gave the iBOT, no matter when it occurred. Tim, having done his duty as a careful CEO, surrendered to enthusiasm. "It's a big bang opportunity," he said. "It will be seen worldwide."

Mike Ferry added to the giddiness by telling them that sales of Reese's Pieces went up 70 percent because of *E.T.* Ray-Bans enjoyed a renaissance because of *Men in Black.* Thanks to *GoldenEye,* BMW took ten thousand preorders of the Z3. Mike didn't mention that BMW reportedly paid $25 million for that placement, or that Mercedes reportedly paid $1 million to place its SUV in Spielberg's *The Lost World: Jurassic Park,* while Dean expected to *be* paid. But why quibble over details when Ginger was about to be discovered by the movies?

That night Dean and I sat in his office library, talking and drinking wine beside bookcases that rose to the high ceiling. A large portrait of Einstein painted by Dean's father looked down from the fireplace.

Dean was low-key. Fred was still a mess. Some DEKA employees thought that the company chosen to manufacture the wheelchair was neither capable nor cost conscious, a combination that alternately enraged and demoralized the Fred team. Dean had given Johnson & Johnson an ultimatum: DEKA needed more control of the project or he was pulling out.

In November, Ginger had lightened his pockets by half a million dollars. "That's serious money." He needed cash fast. Paul Allaire was still a question mark as an investor, but others wanted in at $5 million each. It wasn't nearly enough. He was thinking of selling Teletrol, his climate-controls company, to get six more months of working capital for Ginger. I was reminded of a popular cartoon in which a frog is disappearing down the throat of a heron. The frog, though helpless, is nevertheless strangling the bird. The caption says "Don't *ever* give up."

Dean pushed his hand through his hair, one of his habits when worried, and breathed heavily through his nose. "I used to know all the details of the technologies that we work on," he said. "I wasn't the best on the team but I *knew* about everything. Now, no. That bothers me, because that's the fun stuff. Now I spend my time putting out fires." He blinked slowly. "I wonder whether I'm losing energy." I mentioned that food and sleep did wonders for most people. "I should try that," he said.

He stretched out on the tufted black leather couch, his boots on, wine glass on his stomach. He couldn't get Fred out of his mind. He might put Doug back there to restore order, though that would go off like a bomb at Ginger. But he was desperate. He loved his wheelchair. It was like giving people in prison the keys to set them free.

"I regret that Ginger won't have that kind of impact for several years," he said, "until people get used to it and it catches on. At first it will be seen simply as fun. But that's OK. It's been a long time since I designed just fun stuff."

He pushed his hand through his hair. "Some people just don't understand why DEKA has to change," he said, "even though I keep explaining." His blinks were getting longer and slower. I wished him good night.

The lack of tough questions at the town hall meeting had surprised Tim Adams. He had expected people to ask about stock options. Tim had convinced Dean that the only way to attract experienced people like Mike Ferry and Don Manvel to a start-up was to give them a piece of the action. But the idea of bestowing such incentives on the Ginger team was causing friction among people at DEKA, where Dean had never offered stock. Meanwhile, people on the Ginger team such as Phil LeMay and Ron Reich, as well as one-time DEKA employees such as Doug Field and John Morrell, had been given vague promises about options, but months later still hadn't heard any details. To complicate things further, Dean was insisting that several members of the Ginger team, including J. D. Heinzmann, must remain DEKA employees despite working for the new company. How would such people, or anyone else, share in Ginger?

"It's an explosive issue," said Tim.

His habit when talking was to pace around his office like a caged lion. That's probably the way he felt. He had been at Ginger for eight months without much to do except watch the engineers design the machine. The project's secrecy cramped him, too. But his time was coming. After engineering design, everything else was operational—manufacturing, marketing, supplier relationships, tax strategy, distribution, service, international sales. Those things required experience, which he had by the bucketful and was eager to start using.

At the moment, Tim was devising strategies for dealing with suppliers. Ginger was in a peculiar position with them. The new company had no finished product, no proven market, no bank account in its own name, and no money, yet Dean wanted suppliers to absorb the development and tooling costs associated with furnishing parts for Ginger. "Our business model is unique," said Tim, "because it's them taking the risk." Dean reasoned that once Ginger took off, the suppliers would get rich, too. He was following his usual practice of asking for a lot and shelling out almost nothing.

Some suppliers had other ideas. Why should they absorb the risk? It annoyed Dean when investors treated Ginger like a dot-com start-up, but he seemed deaf to the fact that hardware and tooling were a lot more expensive than bandwidth. To meet Ginger's expected demand for electric motors, for instance, the supplier would need to build another factory. But what if Ginger flopped?

Some of the suppliers were used to dealing with auto or airline companies that guaranteed a certain volume of sales to help offset development and tooling costs. Dean considered volume guarantees unreasonable for a start-up. What if demand for Gingers didn't meet projections? Yet he also expected the suppliers to devote development time and money to the project and to take most of their payment in promises about the future.

The logic may have been contradictory, but oh, that future. If Ginger sold nearly as well as projected, its suppliers would be flush. Take gyros: Each machine needed five of them, so if five hundred thousand Gingers were coming off the assembly line every year, that meant 2.5 million gyros—double the world's current yearly demand, according to Doug. Those numbers were very attractive to gyro-makers such as BAE Systems. If the numbers stood up, Ginger would be four times bigger than that company's current biggest client, said Doug.

Nevertheless, Dean's terms were a tough sell for Tim and Don Manvel, who had been auditioning suppliers to see which ones would invest in Ginger. Even suppliers willing to accept the risk often wanted potentially crippling terms in exchange. That's why Tim, Don, and Doug were on the speakerphone today with Tim's old friend, the former head of procurement for Volkswagen. Tim wanted to hire him to negotiate with suppliers, if he could convince Dean to allow it. The first target was George Sherman, president and CEO of Danaher Corporation, one of whose subsidiaries was making Ginger's motors.

"I want you to protect us from price escalations and from liability for huge volume guarantees," said Tim, pacing around his office while Doug and Don listened attentively. "I want you to check into Danaher's numbers. There's probably more padding because we're a start-up. I think we should be clear that if George wants to build a new plant for this, fine, but we can't be liable for the costs of that. *And* we would be using a second source. I think George is smart enough to know that's sensible."

"And we *would* be their biggest customer," added Doug.

The prospective consultant on the speakerphone asked what George Sherman was like. Tim had met him once and judged him a hard-nosed negotiator, but reasonable. "But if you overlook anything," Tim added, "we're going to get screwed."

A few hours later, Tim, Doug, and Don met with Dean in DEKA's main conference room to update him about suppliers. Doug reported that things were still great with Michelin.

"How much are the tires?" Dean asked.

"$13."

"How much are the cheapest ones you can get?"

"$2."

"If the difference between the best and the worst is only $5 or $10," said Dean, "use the best."

"There will come a time," said Tim, "when saving five *cents* will be important to us."

Though Doug was happy with the design of the motors, Dean wasn't comfortable with a single-source supplier. That was Tim's opening to bring up the supplier consultant. "I don't know how to convince you of this," Tim said, "but we do need help." He added that the consultant was an expert, and therefore expensive on a per-day basis.

"How much?" asked Dean.

"Twenty-four or twenty-five days at $1,500, plus expenses," said Tim. "It could run as much as $50,000."

Dean turned toward Don Manvel, Ginger's director of manufacturing. "I don't mean to put you on the spot," said Dean, "but my gut tells me I'd rather have you do this. You can say $50,000 here and there isn't much, but pretty soon it turns into real money."

Dean also thought it would be a mistake to turn the monitoring and personal relationship over to an outsider. Why not hire the consultant for a day or two, maybe take him to the meeting with George Sherman. But no more. Don looked uncomfortable, as he always did when he disagreed with Dean.

Tim took Don off the hook. "I'll think about everything you've said," said Tim. "Sometimes my experience is a roadblock. But we need a good procurement guy to deal with these suppliers."

"But the big question," said Dean, "and it's the only one, is how much is it going to cost to get this to market? And I don't think we can buy our way to success. I'm not a Puritan who wants to make you suffer, I just don't want you to fall into a pattern that may have served you in the past, when this is a different model."

Tim laughed. "Yes," he said, "it certainly is."

Ginger's current rev was called C1, a sign that the romantic period of invention, the days of Mary Ann and Sybil and Xena, had given way to the practicalities of development. In mid-December, Doug gathered the design team to talk about building the next rev. D1 would be "production intent," meaning that it would be designed as if it were going to be manufactured. This would force the team to freeze as many parts of the design as possible, a crucial step toward production but also a weighty commitment, because suppliers would use frozen specs from D1 to cast dies, tool machines, and start pumping out parts to meet the demands of mass assembly and launch. It could be very expensive if the engineers thawed a part that had been frozen.

All to be expected, except that D1 was a big design leap in terms of almost everything—electronics, chassis, control shaft, transmissions, inertial module (gyros and tilt sensors), software, steering, fenders and wheels, handlebar, and the dashboard displays on top of the handlebar. D1 would be fundamentally different from C1. D1 was scary.

For the D1 prototype, Doug wanted the design team to agree on a deadline of May 2000, just five months ahead. He had written a Mission Statement. The prototype would be "the best machine we can build by May, using the information currently available—no more, and no less."

"No more," explained Doug, meant that if they didn't know how to do something, they wouldn't take a month to figure it out. "No less" meant that if they did know how to do something, they would do it right. He added—and this was the potential kicker—"and we won't leave anything important undone."

He asked for comments. Scott Waters, the industrial designer, was certain he would need to make changes after D1, especially to the control-shaft base.

"Would it give you heartburn if the control shaft didn't change from C1?" someone asked.

"*Definite* heartburn," said Scott.

Mike Martin was the engineer in charge of testing and validation. His job was to torture D1 until pieces of it failed, broke, or wore out. Mike wasn't sure the speedy schedule gave him enough time to beat up D1 and expose its flaws. "My concern is that problems in D1 will stay and become embedded once manufacturing begins."

"But the question right now," said Doug, "is does everyone agree that it's time to get another vehicle integration quickly?"

"And the answer," said Mike, "is *why?*"

"Because complete integration is where you learn the lessons," said John Morrell, the lead controls engineer. "We got burned over and over on Fred because we kept getting fooled by things that didn't work together. Once we integrate, we can choose what to fix."

Doug pointed out that big changes after D1 were a huge risk for tooling. The implied threat of that, noted John, was not to change anything, including things that needed changing. Yet if you made the changes, you risked slipping the schedule.

"You have to believe that if anyone in this room needs wiggle room, we'll find the wiggle room," said John, "because everyone in this room is going to need it at some point."

"But if you need two more weeks for the control shaft," said Doug genially, "that doesn't mean the schedule slips two weeks." In other words, make your changes, but find a way to jam four weeks of work into two.

Doug asked again whether they could commit to the D1 Mission Statement and its deadline.

"Which of the thirty-one days in May does this refer to?" asked J. D.

Doug smiled. "I left that vague on purpose."

"For me," said John, "it's an important test of this group, to see if we can do a machine in five months."

"OK," said Doug, looking around the table, "where should we start?"

That same mid-December evening, Dean piloted his jet to Washington, D.C., with Lucas Merrow, general manager of Fred, and me as passengers. Johnson & Johnson had arranged for Dean to demonstrate the wheelchair the next day at the National Academies.

Dean's news, and therefore his mood, was a bit brighter. He had just convinced Ronald L. Zarrella, president of General Motors North America, to join the board of FIRST. Zarrella was coming to Dean's house for the upcoming kickoff of the FIRST competition, along with actor Noah Wyle, who had recently played Steve Jobs in a TNT production called *Pirates of Silicon Valley*.

The next morning we uncrated the iBOT at the National Academies, backstage of a lecture hall. The presidents of all three academies—Science, Engineering, and Medicine—would be in the small audience. Dean knew them all. William A. Wulf, president of the National Academy of Engineering, was on the board of FIRST.

After opening remarks, someone introduced Dean and the real show began. He rose into balance mode and began tooling around the stage on the iBOT, happy to be showing off this great invention in a place devoted to science and ingenuity. "If anyone in this room had a pushing match with me," said Dean, "you'd lose." Making a balancing machine was easy, he added. Making it *safe* all the time was the hard part.

He mentioned that he had had fifty or sixty people working on this for five years. Lucas Merrow, sitting next to me, clutched his head and groaned, "Oy!" The iBOT, said Dean, was the first real innovation for wheelchair-bound people in decades. "To the wheelchair companies," he said, "R&D is a blowtorch and a hammer."

Then he went into salesman mode, spraying statistics about the number of disabled people, the number of wheelchair users who fall and injure themselves every year, the large percentage of people who move into nursing homes because of incontinence or immobility. "Think how much that costs per year," he said, implying that giving such people a $20,000 iBOT would save the government money in the long run.

Johnson & Johnson had been told that the academy presidents would make brief appearances, but all three were still in their seats, snagged by the machine and Dean's performance. On the way to lunch in the academies' dining room—the presidents were joining Dean—

I mentioned to Lucas that I was surprised Dean hadn't introduced him as the head of the project. Lucas shot me a sardonic look. "That's one of the problems," he said. "To listen to Dean talk, you'd think he made this in his garage."

Dean held the company Christmas party at his mansion every year. It always had a theme, and this year's was medievalism. Some people came as jesters, wenches, peasants, monks, and penitents, but most opted for dressy civilian attire. The Christmas party was the one occasion when some DEKANS perpetrated fashion, wearing sport coats and slacks. A few even wore ties. Dean, of course, wore his usual uniform.

He didn't stint on the party. Food stations were scattered throughout the cavernous house, and waiters circulated with hors d'oeuvres on trays. Dinner was lavish too, prime rib eaten at long tables in the heated hangar where Dean's two helicopters usually sat.

The evening's highlights were the comic skits, filled with inside jokes that tickled the employees and bemused their spouses. The Fred group, for instance, had been pummeling the machine with stress tests. Various pieces had been leaking or breaking, including the machine's arm, which held some of the electronics. A choir of Fred people, led by Luke Merrow, sang a song based on "O Tannenbaum," which included the verse:

Oh upper arm, oh upper arm,
Luke's ulcer is reacting.
Oh upper arm, oh upper arm,
Now he's the one that's cracking.

The main skit always included slides accompanied by commentary in which Dean roasted a couple dozen employees. Although Dean oversaw the skit, it was written and produced by Donna Tamzarian, a longtime DEKA employee whom Dean relied on for textual creativity. She had applied to be DEKA's receptionist back when the company was small and Dean still did all the interviewing. When he asked why she wanted the job, she said she really didn't, but she expected it to be easy and offer lots of time for reading. Exasperated, Dean asked what experience she had. None—she had just graduated with a degree in history—but she didn't think this job could be very hard. Dean couldn't believe his ears, and told her so.

Nevertheless, he hired her, and he remembered the exact phrase that had swayed him. She had described something as "coin of the realm." Dean's theory was that only a smart person would use a phrase like that, and he could always find work for smart people. Now Tamzarian handled payroll and insurance matters and was Dean's all-around copy editor and thesaurus.

This year she had superimposed the faces of DEKA employees onto various medieval costumes and given them comic names such as Sir Bernzalot ("For his skill and expertise in the areas of both deliberate and accidental combustion processes"), the Sheriff of Nodding Off (for the man doing the unending regulatory documentation for the iBOT), and the Royal Pain in the Ass (DEKA's general manager and butt-kicker).

Dean sat on an imposing wooden throne and narrated the show. Every so often, as he called out a comic name, people in black hoods stood up throughout the hangar and shouted, "Stone him!" then threw papier-mâché rocks at the person being mocked. If this mob didn't like one of Dean's jokes, they pelted him. "But I would tell all of you," said Dean regally after one such fusillade, quoting his hero Mel Brooks, "that it is good to be king."

Introducing the Ginger segment of the show, Dean said, "We've got some real *professionals* at DEKA now. They're different from us. One of the best places to get marketing experience is Procter & Gamble. We know this because some very expensive headhunters told us these people could increase sales of diapers and were on the cutting edge of napkins." Howls from the engineers. "So we went from having *no* use for marketers to having *three* of them."

Dean had never hired a marketer and was deeply skeptical of the field. Unlike engineers, marketers didn't make anything. They mostly talked, yet were expensive to hire and had expensive ideas. Dean had started calling Mike Ferry and his two assistants the Three Marketeers or, when he was particularly exasperated, the Three Mouseketeers. For Mike, the joke was already wearing thin.

The DEKA engineers had another Christmas tradition. Two months before the party, about a dozen of them began meeting weekly to design and build Dean's Christmas present. The gift was always elaborate, amusing, and packed with engineering. The engineers treated the project like any other, with frog-kissing, design meetings, deadlines, and prototyping. This year they had created a precisely scaled miniature version of the iBOT—a balancing wheelchair—with a Darth Vader doll as its passenger. The engineers called it a Microbot.

Kurt Heinzmann presented it to Dean and told the party crowd, "Unlike other projects at DEKA, this one had no feature creep." The crowd clapped and hooted. Feature creep was design lingo for what happened when enthusiastic engineers couldn't resist straying from basic design requirements to add a clever upgrade here, an elegant improvement there, and cool features all over the place. The engineers laughed because Dean was a major offender. He found it almost impossible to leave well enough alone. He wandered through DEKA's projects throwing off suggestions like sparks, leaving behind managers frantically stamping out little flare-ups of feature creep.

When Kurt Heinzmann activated the micro-wheelchair and it began to balance, the Darth Vader doll spoke in Dean's voice: "This is *not* a wheelchair. It's an *extraordinary machine*."

Dean laughed with everyone else. "I'm going to bring this into DEKA," he said, "so I can brag when people visit—those poor bastards who can't do what we do."

He loved these moments when DEKA resembled a dynamic, happy-go-lucky family, with himself presiding at the head of the table. There hadn't been many such moments lately, and the future looked cloudy as well.

chapter five

Winter Solstice

December 20, 1999

Two days after the party at Dean's, the forecasters were predicting a white Christmas in Manchester, but Dean wasn't in a holiday mood. Tomorrow, December 21, was the winter solstice, the darkest day of the year. That seemed fitting, especially considering the message he intended to deliver this morning in the annual ritual known as Dean's State of the Union Address.

The entire staff would gather to hear him talk about the past year, the year ahead, and whatever else had been gnawing at him lately. The speech was usually equal parts lecture, oration, pep talk, and homily, all of it seasoned with wisecracks. After the speech, Dean always called each employee's name and gave out Christmas bonus checks, along with handshakes for the men and hugs for the women, what few there were.

But the rumor this year was that there might not be bonuses. Throughout 1999, projects had imploded or turned sour. So as the DEKANS streamed into the big room called Easy Street, they didn't know what might be coming. With Dean you never knew.

His unpredictability was part of the package at DEKA and one of the reasons that people stayed. And left.

A recent hire noticed that many veterans were rolling their office chairs into the room. "Is that a good idea?" he asked.

"Oh yeah. He talks *long*."

Easy Street was about the size of a church social hall. It filled up quickly. Not many years earlier, when DEKA consisted of Dean, a handful of engineers, and a few machinists, everybody could fit into the small lunchroom for these speeches and sit on the floor beside the copier, the Formica tables, and the vending machines. But the company had been growing fast, maybe a little too fast, and now Easy Street was the only space big enough to hold nearly two hundred employees, most of them engineers. Dean no longer recognized all the faces or knew all the names. He had to use a microphone to be heard.

Striding in at 8:05, his dark hair still wet from his morning shower, Dean dropped his scuffed leather briefcase in the middle of the room. After fumbling with the switch on the mike, he stuffed one hand into the pocket of his Levis and lifted his chin like a maestro raising his baton.

"I'll hand out the bonuses in a few minutes," he said softly, retiring that rumor, "but because of some horrible tradition, I'm supposed to talk about where we are and where we're going."

He was leaning against the hulk of a maroon Oldsmobile Cutlass Supreme. Behind him a crash dummy in red pajamas was splayed face-down on the trunk. Engineers used Easy Street to test the iBOT in various environments. There were curbs, steep ramps, a sand pit, walkways made of gravel and loose tiles and wooden slats, and stairs of different materials, depths, and angles of incline. On the room's periphery, spaces were set up like a kitchen, an office, and a living room to test the iBOT's maneuverability. The Oldsmobile sat in the middle of Easy Street because the engineers needed to devise a way to get the wheelchair in and out of cars. So far, that one had stumped them.

"The answer to how we're doing is not simple," continued Dean. "There's some good and some bad."

He began with the bad. Baxter Healthcare, the giant medical company that had been DEKA's main client for a decade, had cancelled two projects. Nor did it bode well for DEKA that Baxter's chairman,

Vernon Loucks, who was one of Dean's father figures and greatest supporters, was retiring this month.

He had more bad news. After many months of work, a DEKA team had finished an ingenious device that eliminated human error in drug dispensing, a problem that cost thousands of lives every year. But the client had run out of money and put the project on indefinite hold, meaning no royalties for DEKA.

"These things have left thirty or forty people here with no project to work on," said Dean. "But my rule is that no one will lose a job for causes that aren't their ethical or professional fault. The consequence is that DEKA—that is, me—had to absorb that expense."

There were also problems on the iBOT project—more delays and more troubles with Johnson & Johnson. Dean pushed his hand back through his thick hair, as if to soothe himself.

But there were bright spots. The Stirling engine project, for instance. Dean called the Stirling engine the golden fleece for engineers, because the theory behind it was so tantalizing, yet the machine itself was so difficult to realize. Internal combustion engines typically convert only about 25 percent of fuel into energy, wasting the rest through exhaust, heat loss, and other factors. By contrast, a Stirling engine was theoretically almost 100 percent efficient. It used and reused a fixed amount of gas in a closed cycle by constantly transferring the gas between a cold end and a hot end, which could be heated by anything that burned. The gas's expansions and contractions drove a piston.

But no one had been able to build one that was simultaneously reliable, practical, manufacturable, and affordable, despite hundreds of attempts since a Scotsman named Robert Stirling had filed a patent for it in 1816. But Dean was pretty sure he had figured out the problem, and DEKA's Stirling team seemed to be confirming his idea. If the project succeeded, it could transform the energy system and provide cheap power to the Third World. "Even a moderate success," Dean told Easy Street, "would overwhelm what we've made before."

His report on Ginger came next. He finally smiled. Most people in Easy Street knew little or nothing about the machine. "It's probably the largest project we've ever had," said Dean. "Conceiving, designing, and prototyping new ideas is what's fun, and it's what we do. Tooling, inventory, manufacturing, and service we have tried to avoid. We've let the big boys do that. We go along for the ride and collect a royalty. But there is no obvious big guy in the Ginger business, because there is no product like Ginger. So we've decided to be the captain of our own soul and do Ginger ourselves."

He ran a hand through his hair. "The trouble is that I still have this bad habit of not wanting to be at the mercy of someone else's incompetence."

That led into one of his recent favorite topics for a rant, venture capitalists, whom he often called vulture capitalists. Venture capitalists wanted to feast on Ginger—on *him*. They epitomized the things that disturbed him about the New Economy. They lived off the ideas and creativity of others, contributing nothing but money yet demanding control. They weren't drawn to a project by its ingenious engineering or its prospects for changing the world, but by that most mundane of motives, the chance for a quick financial turnaround. Too many companies these days were selling their creative ideas into VC slavery, betraying themselves and their employees for a few pieces of silver. If he could, Dean would have retained absolute control over Ginger by funding it himself. Since that was impossible, he would hold his nose and deal with investors, but he was determined to do so on his own terms or go broke trying. Which was exactly what was happening.

"Venture capitalists love to see you slip," said Dean. "That's how they get control—and a Faustian deal with the devil has never seemed smart to me. But at the same time, Ginger needed to move along, so we've continued to fund it, and there's now a substantial number of people. It's *very expensive*," he said. "And in recent days I've gotten some surprises. It's going to cost more than we expected. The marketing plan needs more money than we thought. The tooling costs are more than we expected. So if we had signed with a venture capitalist, we would be on that slippery slope."

He looked at the floor as if it might suddenly tilt away from him. He didn't tell them that he had spent $500,000 of his own money on Ginger just in the past month.

"In March I had my back cut open," he went on, referring to disc surgery he'd had nine months ago. The reference seemed to come out of nowhere, like other personal details sprinkled into the speech: "A number of people have noticed my lack of food and sleep this year" or "People have been telling me to take better care of myself." At first such comments seemed odd and irrelevant, until you realized that Dean didn't differentiate between DEKA's health and his own. His worries about one were connected to the other. The company was a projection of his brain and heart, his personality and goals, his genius. It ran on his money and his ideas for inventions. He didn't have a wife or children and spent little attention on his girlfriend, K. C. Connors. He called DEKA his family, and he ruled it like a strict and usually benevolent patriarch. Sometimes he had to deny them things they thought they wanted.

"The world is changing," he said. "If we're going to compete for the people we need, compete with these dot-com companies that give people a lottery ticket mentality, it's hard, because we don't have some of the currently attractive tools. But I can't do things just because others do them if I think they're fundamentally wrong."

Currently attractive tools? He couldn't bring himself to say the hated words that attached themselves to venture capitalists like remora fish to sharks: stock options. Even Mike Ambrogi, his friend and right hand, had been pressuring him to get with the nineties and give DEKA's employees a greater financial stake in the company. Mike had joined DEKA eight years ago. He remembered thinking during his interview that Dean was either full of shit or the most amazing guy he'd ever met, because he had so many ideas about changing the way people look at things. When Mike had asked what his job would be, Dean had said, "I don't know. You'll do what you're good at."

Mike signed on. He turned out to be good at managing people and projects. He was the mechanic who put in sixty-hour weeks to keep DEKA operating no matter how hot Dean ran the engine. He called himself "the one person at DEKA who worries." He was steady, demanding, and good-natured, and was universally liked. He was also one of Dean's few real friends, the sort Dean could drop in on and share a meal with.

But as Mike watched the Internet revolution pass him by, watched MIT classmates with less talent become dot-com millionaires, watched Dean buy a Humvee and a Citation jet and build a 32,000-square-foot palace on a hill, he had gotten restless. He had helped build DEKA and he wanted Dean to cut him in. He was thirty-six, with young children. His wife was urging him to make his move.

Other people at DEKA were feeling the same way. In many companies, the people who created new products received founder's options, but at DEKA the engineers who turned Dean's ideas into wonderful machines just kept collecting their salaries. Dean got the royalties.

Dean's aversion to stock options was making it harder to entice the best engineers to DEKA. He had always low-balled salaries, preferring to attract top prospects with great projects and his own charisma rather than with great pay. But by the end of the 1990s there were promising new projects everywhere. Fresh graduates and many experienced engineers saw their future in Internet technologies, and if they had to settle for low pay, they expected sophisticated incentives.

But Dean kept running DEKA the way he always had. His mother still did the books, for crying out loud. This folksiness was once part of DEKA's nonconformist appeal for longtime employees like Mike. Now it just seemed antiquated.

Mike had tried talking to Dean about all this many times, but Dean didn't want to hear it. DEKA was all his and that's the way he liked it, partly for financial reasons, but mostly because it gave him complete control over which projects he funded and how much he spent on them. People with stock options might expect him to heed their opinions.

Sycamore Networks, a Massachusetts company that designed optical networking, had offered Mike an attractive job in April, but he had said no, out of loyalty to Dean. Six months later, the options that had been part of the offer were worth $10 million. Mike couldn't stop thinking about that. The iBOT's chief systems engineer had left for Sycamore a few months ago and had become a paper millionaire within months.

Mike loved DEKA, loved its energy and talent and the excitement of trying to keep up with Dean. But by autumn of 1999, he had had enough. He called Sycamore to see if they were still interested, and when they enthusiastically offered him a plum job, director of program management, he took it. On October 15 he had walked into Dean's office and told him that he was starting at Sycamore in January. Dean was stunned. Mike said he was tired of doing middle-management scut work with no stake in the company.

But this is my *family*, said Dean.

But it's not *my* family, said Mike. I need to go out on my own and see what I can do.

Dean offered him more money, but that wasn't Mike's point. A bigger salary was still a cap. Dean reminded Mike that he also got an annual bonus, but to Mike that was just another purse string controlled by Dean. He was tired of depending on Dean's variable largesse. He wanted an unlimited upside, a direct link between a company's success and his own. He couldn't have that at DEKA.

In the next weeks, Mike watched Dean go through stages. First, denial. For a month he refused to talk about transitional matters related to the departure. When Mike pressed him, Dean would sit behind his desk with his head in his hands, not speaking for minutes at a time. It was weird, seeing Dean like that, quiet and stricken.

Mike anticipated the next stage—anger. Dean called him a quitter, tried to make him feel terrible, said he was selfish, that he was abandoning him at the worst possible time, when so many things were going badly for DEKA. It was almost impossible to rile Mike, but this time he bit back. *He* was being selfish? After devoting his heart and soul to Dean for eight and a half years? For a *salary*? He heatedly said that maybe he should quit right then.

That calmed Dean down. He entered a new phase, wistful and reasonable. He told Mike that he wanted to have more fun, do more engineering again instead of putting out fires and schmoozing investors. He asked Mike to hang around for just a couple more years.

A couple more years! Mike shook his head. As a last resort Dean held out a carrot. He expected DEKA to spin off new companies within the next few years. Mike could manage one of them and own a small piece. For Mike, it was too little too late. "I'd still be his assistant," he thought, "and he casts a *big* shadow." The chance to grow in his own light exhilarated him.

Dean's last phase was sad resignation. Mike hoped it lasted. Like other charismatic leaders, Dean sometimes acted as if people who left DEKA had somehow betrayed him. Mike wanted to avoid that. Dean was one of the most remarkable people he'd ever met, and he hoped to preserve their friendship. So did Dean. If I have to choose between working with you or having you as a friend, he told Mike, I'd rather have you as a friend, because I don't have that many.

When Mike announced his departure just before the Christmas party, people were shocked. Dean was DEKA's magnetic visionary, but Mike was its rock. When other projects had been crashing around them this year, people had told themselves, "Things must still be OK, because Mike's here."

Now, on Easy Street, Dean's thoughts jumped from VCs to dot-com rapacity to Mike's departure.

"We're a company where management doesn't drive things, judgment does," he said. "Mike brings judgment. Most big companies are run by policies. We don't have policies, we have a place with *soul*. Mike represents that soul. My hope is he'll go, make a lot of money, and come back. Mike's a person you can always depend on." He turned to look straight at him. "That's why I think you'll be back."

Mike's departure was the year's nadir. Time to rally the troops. Dean pulled a crib sheet from his shirt pocket. "But the upside," he said, "is that even though pressures and anxieties are higher, opportunities have never been greater."

For example, Johnson & Johnson expected to charge between $20,000 and $25,000 for the iBOT, and the market worldwide was easily 50,000 to 100,000 machines. Dean pointed out that even if the iBOT was a relative failure and sold only 30,000 machines, DEKA's cut would be tens of millions of dollars.

DEKA was also building disposables for a new company's hemodialysis machine and working with a company on a machine to treat scleroderma, a rare blood disease. Dean expected Baxter to return with its tail between its legs for help on two projects for which DEKA had the expertise. Dean had recently made another potentially lucrative deal with Johnson & Johnson to develop a drug-delivery device.

He looked around Easy Street. "So despite everything that's happened this year, all the setbacks, everybody is still at DEKA and the technology is still *ours,* and we may see it used in the coming year."

This thought reminded him of Ginger again, the project most on his mind. And that reminded him of Jack Hennessy from Credit Suisse First Boston, and of Michael Schmertzler, who worked for Hennessy as cohead of Private Equity in the United States. Dean told Easy Street that CSFB seemed very interested in Ginger. Schmertzler had even come from New York City to attend DEKA's Christmas party and had stayed at Dean's late into Sunday. Schmertzler was a tough poker player, but he couldn't hide his enthusiasm for Ginger.

"It's not a deal until it's a deal," said Dean, "but it looks promising." He was emphasizing the optimistic for the sake of his listeners. Hennessy and Schmertzler were still playing coy, but DEKA didn't need to hear that. "My idea of a good partner for Ginger," Dean added, "is someone who leaves us in total control. We've spent a few million this year waiting for the right partner."

It was time for another turn in the narrative, toward the finish. "Unlike you," he told them, "I don't have the option of leaving. My father has called me a 'human irritant' since I was four. He said that I'll always do anything I decide to do. I *have* to keep these projects going. I *won't* give up."

So far his soliloquy had lasted two hours. People occasionally yawned, looked around, shifted their weight, daydreamed. For some of them, DEKA was a job, not a mission.

"But it wasn't a good year for profits," Dean was saying. "We don't pay people overtime and we don't charge clients for overtime. I don't believe in that. So how do you compensate people for working extra hours? You give them a bonus. How do you compensate people who could make more elsewhere? You give them a bonus."

The bonus pool had always been 15 to 20 percent of DEKA's royalties, and people expected the bonus to swell their salary by roughly the same percentage. But what about the years when losses outweighed royalties? Some companies offered stock in such circumstances, but not DEKA. And people should be glad for that, Dean said, because stockholders always tightened the noose and people lost jobs.

He was still leaning against the wrecked Oldsmobile, his boots crossed. He talked softly into the mike, as if he were thinking out loud at a family get-together.

"Less than 30 percent of the people in this room have worked on a project that's bringing in royalties. I try to be fair. Fairness is a complicated concept. Some people want it both ways: 'My project didn't work out but I want a bonus anyway.' "

He shook his head a little. He asked them to do a mental experiment: Imagine DEKA a year from now, when projects will be humming along, generating millions in royalties. "Because whether you like it or not, I am your 401(k) manager. I'm investing in Ginger *for* you. Imagine what life will be like when we can fund internal projects and I can eat and sleep again."

He ran a hand through his hair. "I don't think it's hard to imagine." They would have to trust his imagination, his business moves, his plans for them. He stood up.

He was almost to the end of his story, to the only part that some of them really cared about: How much would their bonuses be? Last year, when things were better, the pool for bonuses had been just under a million dollars. He implied that that was impossible this year. "I expect the pool for bonuses to be enormous in the future," he continued, offering a promissory note, "and the percentage donated to that pool won't change." He predicted that the pool would double next year and be five or ten times as big within a few years. But he added that not everyone would get ten times the amount of this year's check. As always, he would decide who deserved what.

"I'm tired of hearing about stock options," he said, still scolding. "We share when it's good and we don't when it isn't. You either buy into that or you don't. If you don't," he said, looking around, "I'd like to know soon. I want people who share the vision and soul of this place."

He finally told them that he had decided to keep the bonus pool the same size as last year's, just under a million dollars, but there were more people splitting it. In a tight voice, he added that he had gotten the bonus money by mortgaging his paid-for palace. He looked grumpy. He hated owing money. *Hated* it.

"So to the people grumbling, I want you to stop," he said in a chiding tone. "To the people not enthusiastic, I want you to get over it. And the people who want to sleep on a bearskin rug had better make sure the bear is dead first."

He acted like a father who was going to give his teenager an allowance, but not before lecturing him on his attitude and shortcomings. "I'm *going* to do these projects," he said again. "I believe the world needs them all, whether it's FIRST or Fred or Ginger or the Stirling. I *will* make them work."

It was his summation, but he still wasn't quite ready to hand over the money. "We can't keep going the way we have been on some of these projects." Everyone knew he meant Fred and Ginger. He pointed to his briefcase, where the checks were. "I'm doing this because some of you have worked hard and because, believe it or not, I enjoy sharing success and giving gifts. And a portion of this *is* nothing less than a gift. I'm going out on the biggest limb of all—debt—which gives me the right to say to you that there are people who *do* it and people who *watch*."

The room was silent. The sermon seemed to be over. He had been talking for two hours and forty minutes.

Mike Ambrogi stood up from his bleacher seat. "Can I say something?"

"No," said Dean, half kidding.

"Come *on,*" said Mike. He took the microphone from Dean but couldn't go far because the power-pack was clipped to Dean's belt. "Attached at the hip to Dean," he quipped. "How novel."

Mike looked around the room. "I want to thank Dean for teaching me, for being a mentor and a friend," he said. Before he reached the end of the sentence his voice started trembling and his eyes filled up. "*Shit!*" he said, and took a deep breath. "It's been an honor to be part of this group." Overcome, he hugged Dean, who looked startled but awkwardly hugged him back.

Easy Street applauded as if they had just watched the final scene in a play, but there were many acts to come.

chapter six

Lions and Angels

January–March 2000

Dean was glad to see 1999, which he called the worst year of his life, come to a close. The financial burden of Ginger was worsening, with no relief in sight. The group of potential individual investors had disintegrated. Credit Suisse First Boston was still making kissy noises, but kept hedging, trying to get more ownership and refusing to put anything on paper. As the new year opened, Dean thought he might have to sell a large chunk of the Stirling engine project to investors or borrow money against future dialysis royalties. He was depressed and desperate. That was exactly where Jack Hennessy and Michael Schmertzler of CSFB wanted him. The longer CSFB could stall, the better terms they could extract. Dean always led them to believe that he was holding aces—as the old poker adage puts it, chicken today, feathers tomorrow—but the bluff was getting harder to sustain.

He called his old friend, Vernon Loucks, the recently retired CEO of Baxter Healthcare, to see if he or anyone else might want to invest in Ginger. Loucks soon corralled a couple of prospects. Dean also brought in William A. Sahlman, a professor of entrepreneurial studies at Harvard Business School, who had extensive contacts. Nine rich men soon made plans to travel to New Hampshire for a meeting at Dean's house during the last week in January 2000.

Dean told two different versions of what happened next. In the first, he raised CSFB's bet by informing Jack Hennessy that he was about to host an eager group of billionaire investors. Hennessy told Dean he couldn't do that and hopped on a plane to New Hampshire the next day. You're crazy! he told Dean. We're giving you the money, and that means *we* decide all sorts of things.

It's the same deal as it was a month ago, replied Dean, the same deal as it was two weeks ago, the same deal as it was yesterday. So it may be crazy, said Dean, but it's consistent. Hennessy returned to New York and quickly drew up some papers, which arrived at Dean's house by Federal Express on the very day he was meeting with the new group of potential investors.

In the other version, Hennessy didn't know about the meeting until he happened to call Dean while it was in progress. When Dean cheerfully told him what was going on, Hennessy went nuts and flew to Manchester the next day and agreed to general terms.

Whichever version was true—Dean didn't let details get in the way of his stories, which usually starred himself—Loucks's and Sahlman's investors did come to Dean's house, listened to his spiel, rode Ginger, and began salivating. Five of the nine, according to Dean, were billionaires. Afterwards, the group sat around Dean's dining room table, with Bill Sahlman moderating. One potential investor, whose fortune came from broadcast and cable businesses, said he was in for $10 million if Dean would give it to him, or $5 million plus whatever was left. The guy next to him, from Shanghai, wanted in as well. Others around the table started making offers, too, until Sahlman stopped it as premature.

Dean was elated. Such investors, he thought, are aptly called angels. He thought he might raise $50 million—$10 or $20 million from the private investors, the rest from CSFB. "For a single-digit interest in the company," he pointed out to me. "Not even the dot-com guys can do that. It also means we wouldn't have to go back to the investors for more money, because the $50 million would carry us through to launch."

One month into the year 2000, it looked like his luck was about to change.

During the first week in February, I met Dean and Lucas Merrow, manager of the iBOT project, at the Michelangelo Hotel in midtown Manhattan. Johnson & Johnson wanted Dean and other people involved with the iBOT to get some media training. The chief instructor was a stylish, carefully put-together woman with blood-red lipstick who used phrases like "from an outlook perspective" and "I'm asking for a mandate." Her business was to teach executives how to stay on message and put the best spin on everything. Today she interviewed each participant on tape, then did critiques.

She managed to stay on message despite Dean, who acted like a one-man peanut gallery, constantly quipping about journalists' imbecilic questions. "Why didn't you ask her what *her* salary is?" he said. Or, "At a moment like that, you bite your lower lip and say, 'I love you, Hillary.'" During his taped interview, when he was asked if the iBOT wasn't really just another wheelchair, Dean said, "This is like a wheelchair the way a pigeon is like the Internet."

Part of the day's work was to train people to answer certain questions according to corporate dogma. The Johnson & Johnson people were still wincing from Jane Pauley's closing comment on *Dateline*: "Brace yourself for the price—$20,000." They wanted to discuss a strategy for answering the inevitable question, "How much will the iBOT cost?" Dean listened to various watery answers, then interrupted. "The answer is, 'It cost nine years and more than $100 million in research and development to create this device that serves a small disabled population.'"

Afterwards he said, "They seemed to be coming at it from the wrong direction: What are the answers people want to hear and how can we make that consistent with the truth? They had terrible answers to some very basic questions, like 'Do you expect to make a lot of money?' What's wrong with saying, 'Yes, but it's going to be a long time before we start making a profit after spending $100 million to liberate people, when the government wouldn't do it and other companies wouldn't do it.' They give plastic answers when they should be excited and proud."

The financing for Ginger kept hitting barriers. Once Credit Suisse First Boston finally committed, the bank didn't want to share the investment with the private group. Michael Schmertzler, cohead of CSFB's Private Equity division, had taken over the negotiations, and he was adamant: His group wanted all $50 million. But Dean couldn't do that to his friends Vern Loucks and Paul Allaire (who was back in), and to *their* friends. He liked his angels, who incidentally gave him leverage with Credit Suisse. By mid-February Schmertzler had agreed to $40 million for CSFB, with $10 million spread among the private investors.

Schmertzler drew up papers and sent them to Bob Tuttle, Dean's financial expert. For twenty-five years, Tuttle had negotiated all of Dean's big deals, and was his most trusted advisor. Tuttle's office was just down the hall from Dean's. Everyone at DEKA knew who he was, but few people had ever heard his voice and even fewer had seen him smile. Tall and slightly round-shouldered, with dark hair that fell across his forehead, he sat in meetings like a phantom with a poker face, rarely speaking. He was calm to the point of impassivity. If someone set his hair on fire, he would consider all his options before jumping up. Dean had never seen him either angry or exuberant, and jokingly described him as silicon-based. But he admired Tuttle's gifts and accepted his advice. "I've seen him go into a meeting where five lawyers are trying to get things from him," Dean once said, "and we *always* win."

On February 14, as Tuttle went through Schmertzler's term sheet, Dean flew to New York. Schmertzler had invited him to the Private Equity Division's board meeting the next morning at 8:00, where he wanted to announce the deal while Dean demo'ed Ginger. Dean was in high spirits. When we landed there was a message from Tuttle on Dean's cell phone: CSFB is moving sideways. Dean called Tuttle back on the ride into the city. Tuttle didn't like what he had found in the documents. Schmertzler had tried to sneak in new terms to CSFB's advantage. Tuttle had been tussling with him on the phone for the past two hours.

Schmertzler was a worthy opponent with a stellar résumé, starting with an M.B.A. from Harvard. He had been head of international investment banking and other departments at Shearson Lehman American Express. At Morgan Stanley he had served as president of the firm's Leveraged Capital Funds and Leveraged Equity Fund II, and as a managing director in the departments of Corporate Finance and Mergers and Acquisitions. From Morgan he had moved to his position at CSFB. Dean called him the sharpest financial guy he had ever seen—except for Bob Tuttle.

Neither Tuttle nor Schmertzler had given an inch in the phone conversation. Dean looked irritated when he hung up. He saw three options: (1) Schmertzler could change his position; (2) Dean could call Schmertzler's boss, Jack Hennessy, and ask him to step in; or (3) Dean could call in the angels and offer them CSFB's share. He wasn't desperate anymore. He felt confident he could fund Ginger one way or another. It looked to him like Schmertzler had outsmarted himself. Tomorrow morning Ginger would dazzle the board members, but then Schmertzler would have to admit that he had gathered them prematurely. The deal wasn't settled.

The next morning Dean lugged a heavy duffel bag into a conference room in CSFB's building on Madison Avenue. Schmertzler greeted him. Short and trim, Schmertzler was an intensely observant man who rarely blinked. He kept flicking his vigilant eyes at me, first because he didn't know who had entered his kingdom, and then because he did.

He asked Dean if he had talked to Tuttle. Dean casually said that he'd gotten a message that Tuttle wasn't happy because Schmertzler had put in terms they couldn't accept. He added, even more casually, that Vern Loucks and the billionaire angels had become importunate, wondering when they could send him their money. Schmertzler's face mixed pique and grudging admiration. First Tuttle had foiled his ambush, and then he had fallen into Dean's trap. It didn't happen often.

About a dozen people sat around a grand polished table. Schmertzler showed the *Dateline* tape, then Dean put in the Ginger tape. While it was running, he unzipped the duffel and began assembling a Ginger. When the video ended he started scooting around the conference room, pirouetting and talking without pause about all the markets for the machine: malls, airports, theme parks, pedestrians, security guards, golfers, mail carriers, delivery people, warehouse workers. "People haven't realized that something needs to do to the car what the PC did to the mainframe."

"If I want to take it to Smith & Wollensky for lunch," asked one beefy board member, "would it be safe outside?" Yes, there was a security code as well as protections in the software that rendered stolen Gingers useless.

Price? Dean asked how much *they* thought it would cost. Someone guessed $15,000. Dean smiled—he said he could make them for much less than that.

Patent protection? Another smile. "You can't get any broader than our patent," he said. He told them how he had gone to the Patent Office to argue with an inspector who didn't want to grant such a broad patent. After two hours, the Commissioner of Patents, Q. Todd Dickinson, had entered the room and asked if there was a problem. Dean rolled over to him on Ginger. "*How* did you do that?" asked Dickinson. Dean explained, then asked Dickinson to show him something similar or give him the patent. He flashed another smile at the board members. "I got the patent."

He had sold them, no question. The Smith & Wollensky guy wanted to ride. He learned quickly, but was overconfident and crashed into the coffee bar. But he was wearing that Ginger grin. He wanted in.

Dean left the room feeling jubilant. The year 2000 kept looking more and more promising. Next week he was flying to Acapulco to speak to a convention of pneumatics suppliers, because the director had guaranteed sponsorships for more than two hundred FIRST teams. Then he was going to California to speak at the TED (Technology Entertainment Design) conference, a gathering of CEOs and other influential people whom Dean thought of as "low-hanging fruit" waiting to be harvested for FIRST. He might make a side trip to talk to Spielberg's people again. And now the financing for Ginger seemed all but sewn up.

Dean flew his Citation jet twenty-four hundred miles to Acapulco and then another two thousand to the TED conference in Monterey, California. The Citation was a small twin turbofan whose cabin seated five. It cost about $5 million, plus $500,000 per year to operate and maintain. Dean's full-time aviation mechanic took care of it. Dean was one of the few people in the country certified to fly such a jet without a copilot. To keep his certification he had to pass an annual battery of rigorous tests in a flight simulator. Failing any phase meant decertification. He never had time to prepare for the tests but always passed.

He hated to fly commercial and only did so for overseas trips. With the Citation, he could leave on his own schedule, without the friction of check-ins, layovers, and cramped seating. Because his pockets served as tool chests, just getting through the metal detector was a hassle.

The biggest problem was his little Swiss Army knife, an essential item. Even before the terrors of September 11, 2001, the inspectors insisted that he give up that tiny blade. So he developed a strategy. While they were clucking about his little knife, he would tell them not to worry about that one, because he kept a much bigger one in his briefcase. Naturally they demanded to see it. And there, balanced in one of the briefcase's pen loops, was a typical steel dinner knife stamped United or American. Sir, the inspectors would say, you *can't* take *that* on the plane.

Dean enjoyed that moment. Fine, he would answer, as long as you can explain something to me: Twenty minutes after you confiscate this knife, we're going to be in the air and the attendants are going to serve dinner—and they're going to hand me a knife just like this one. Why is that knife OK but this one isn't?

In those days before September 11, the inspectors looked befuddled and then let him through. Every time. So he always kept a butter knife in his briefcase to help him get his little Swiss Army knife through.

At the TED conference the previous year, he had spoken about FIRST. This year the organizers wanted him to talk about the iBOT. Dean rode one onto the stage in balance mode, worked the crowd, then rolled down the stairs and out, amidst cheers.

There waiting for him was John Doerr, the kingmaker of Silicon Valley venture capitalism. A partner at Kleiner Perkins Caufield & Byers, Doerr had bankrolled Compaq, Netscape, Amazon, Sun Microsystems, and other huge New Economy successes. His name made dot-commers stutter and sweat. Even Dean had heard of him. Doerr told Dean that Bill Sahlman, the procurer of angels, thought they should talk. Instead, Dean handed him a packet of literature about FIRST. "Here I am giving homework to the god of venture capital," he recalled, grinning.

Doerr was passionate about education, and when they met the next day he seemed excited about FIRST. Dean decided, as he later put it, to go for the jugular. He asked this legend whom he'd barely met to become FIRST's West Coast director. Doerr paused, surprised by chutzpah that matched his own, then agreed to think about it. But he also wondered what Sahlman wanted him to see. For that, said Dean, Doerr would have to visit New Hampshire.

Meanwhile, Jack Hennessy and Michael Schmertzler were frantic at Credit Suisse First Boston, convinced that Dean was trolling for investors in Silicon Valley. Schmertzler left a number of messages for Dean: I'm upset, Hennessy is upset, we thought we had a deal, let us fix whatever's wrong. Hennessy left similar messages and offered to fly to California the next day to sign the deal by lunch. Dean had his assistant tell Hennessy not to bother. Schmertzler phoned again, sounding agitated, to say that he and Hennessy had their plane tickets but wouldn't be coming, since Dean wouldn't return their calls. He offered to meet Dean in California, Utah, or anywhere else. He blamed the misunderstanding on things inserted by CSFB's lawyers and pleaded with Dean to let him fix the deal.

Dean was sure that Schmertzler had given the lawyers their instructions, but Bob Tuttle had caught him and now Hennessy was apoplectic. The most bothersome clause guaranteed CSFB a timely $400 million return on its $40 million investment; otherwise the bank got a bigger chunk of stock.

Dean and Tuttle played it masterfully. While Dean stayed maddeningly out of touch, harvesting kudos and who knew what else among the billionaires of Silicon Valley, Tuttle deflected all of CSFB's counterproposals. Schmertzler, outmaneuvered, called Tuttle at the end of the week and agreed to do it Dean's way.

But by then the momentum of the game had changed and Dean wanted a different deal. Tuttle told Schmertzler that Dean needed the weekend to think things over.

That weekend, Dean and Bob Tuttle reconfigured the Ginger financing. They cut CSFB from $40 million to $30 million, with Sahlman's angels divvying up the other $20 million. Bob Tuttle had given Michael Schmertzler this news on Monday, February 28, and was waiting for a response.

On Tuesday morning, Dean and some members of the Stirling engine project were meeting with representatives from Johnson &

Johnson and the management consulting firm McKinsey & Company. DEKA and Johnson & Johnson had hired McKinsey to study the Stirling's competitors and possible markets. Johnson & Johnson had funded a portion of the project's preliminary phases, but the time had come to push in big money or bow out. Dean wasn't worried. He knew a few places to find investors.

Dean had just finished telling the Johnson & Johnson people that his little Stirling could power a village in India and distill drinking water as a byproduct when Tuttle came in and whispered into his ear.

"So he's agreed to *everything*?" said Dean, sitting up straighter.

"Yes," said Tuttle.

Dean glanced at me with a half-smile. Hennessy had promised to send a term sheet by tomorrow afternoon. He wanted Dean to come to New York next week to close the deal. Dean looked happy for a moment, then began gnawing on his thumb.

The next day Dean, Tuttle, and Tim Adams met with another potential angel, Robert Halperin, the former president of the Raychem Corporation, a large electrical products company. Halperin's folksy, grandfatherly manner contrasted with his flashy pinstripe suit. He asked pointed questions and gave blunt opinions, some of which Dean liked, such as his advice to launch with a high price and to manufacture millions of Gingers right away. He told Dean not to worry about paying his investors back quickly, that his first priority was to capture the market and box out competitors.

But some of Halperin's advice was jarring. He asked why Dean was determined to keep 80 percent of the company for himself. "Because if I give up 20 percent to the investors and 10 percent to insiders," said Dean, "and then need another round of investment, I'm below 50 percent. And I sleep better having 51 percent than 49 percent."

"In a word," said Halperin, "that's absolutely ridiculous." Dean looked taken aback. "Your main concern shouldn't be your percentage," continued Halperin. "You need to run the company well, and you can do that with 1 percent."

Afterward, Dean gave him a lift to Boston's Logan Airport in his helicopter. "Something is going to go wrong," said Halperin. "I have no idea what, or what form it will take, but it *will* happen. So you need people who know how to help when the unexpected happens." People, he implied, like him.

As Boston's saffron lights came into view, Halperin offered to put in $5 million now and up to $100 million in the next round. We left him at the airport and lifted off toward Manchester. Below us a stream of red and white lights inched along Boston's congested highways.

"He offered up to $100 million," said Dean, stunned. "He doesn't care about the money. There is nothing these guys want that they don't have. They want to do something *interesting*. That's why they look for something like Ginger to invest in. He didn't even ask what his return would be, or when."

Halperin's crack about the percentage of ownership was still nettling Dean. "People think it's so I can walk into a board of directors' meeting and say, 'We're doing it *this* way.' They don't understand. I like it to be *my company*. It's not about the money, either. If I do something that costs $20 million and it's a mistake, I want to be able to take $20 million out of my company and return it to the investors, and you can't do that if it's not your company."

We flew in silence toward the dark hills of New Hampshire. "Johnson & Johnson is not Ralph Larsen's company," Dean continued. "Baxter is not Vern Louck's company. But DEKA *is* my company. The people there are *my* people. These guys wouldn't understand that."

If DEKA wasn't his, he went on, he couldn't have sunk all that money into the early stages of Fred and Ginger. He couldn't have celebrated the launch of his dialysis machine by chartering a 727 and taking the whole company, spouses and kids included, to Disneyland, the San Diego Zoo, and a movie studio. Some of his people had never been on a plane before that. Some had never been out of New Hampshire. He took care of his people. To do that he needed control, and that meant ownership.

I once asked him why he had never married. "I guess because the reason you get married is to have kids," he said, "and if I had kids and did it right, I couldn't have DEKA and Teletrol and Enstrom. So if I did it right, I'd resent the kids, and if I didn't do it right, I'd feel so guilty and terrible. So I decided not to. I have a hundred and sixty kids at DEKA to take care of."

He sincerely felt these things: the entrepreneur's need to stay independent of investors, Papa Dean's need to do right by his people, the inventor's need to devote himself to innovation and business without the distractions of a wife and children. Yet beneath these noble motives lurked another that was more self-serving and potentially damaging: the need for absolute control, and an obsessive fear of losing it. This, too, guided Dean's business decisions and in some ways isolated him personally. His acquaintanceship was vast, but he had few close friends. He had been living with K. C. Connors for six years, but treated her more like a personal assistant than a lover. "Dean's great at everything except relationships," K. C. once remarked to me. Love entails loss of control, sharing of power, even occasional submission, none of which Dean would willingly allow. Whatever the situation, he always needed a large controlling percentage. Maybe it was good to be king, but as Shakespeare's monarchs illustrate so eloquently, it could also be a cage.

Meanwhile, Ginger's engineers had other worries. At the daily morning briefing the next day, the team heard that Doug Field was at the hospital with an accident victim. The marketers, desperate for consumer data, had gotten a DEKA spouse to ride Ginger as a "customer surrogate." When she went over a curb, the machine momentarily lost traction, and when the wheels caught again, the jolt threw her off backwards. She hit the back of her head, which pushed the helmet onto her glasses, cutting and bruising her nose. The marketers wanted details about what had happened; the engineers wanted details about why.

Jon Pompa, a young engineer and dirt biker with bleached hair and an earring, had started working on the problem. When one wheel was slipping but the other had traction, Ginger got confused by the mixed message. When both wheels were slipping and then suddenly found traction, the rider could be thrown backwards. And when Ginger went off a jump, putting both wheels into the air, a rider would normally lean back, which signaled the processors to order the wheels into reverse. And that's where they tried to go when they first hit the ground, despite the machine's strong forward momentum. By that point the rider was leaning forward, too, which told the machine to accelerate ahead. The contradictory combination of commands and momentum caused the machine to buck violently and could throw the rider off.

Jon was trying to adjust Ginger's behavior via the controller. "The machine gets mushy at the long end of the wobble," he explained, "but if you try to make it squeal, it won't shoot out from under you and give you unhappy oscillations." The engineers nodded.

Ginger also needed to lose weight. Lighter was better in terms of range, acceleration, and appearance. The original specification was 65 to 70 pounds, but Ginger had gained more than 8 pounds since the last revision and C1 weighed almost 90, with more heft sure to come. Doug had asked the team to shave 10 pounds for D1, the rev they were working on now. He wanted the weight spec to begin with a seven. "So 79-point-something pounds," said Ron Reich.

They went through the machine part by part, looking for slivers of fat. The first few pounds were easy. The second few were tougher. Then the real work started, on the last 4 pounds. Most of it had to come from Ron Reich's chassis and Bill Arling's control shaft (the vertical part of the T-bar). The two of them studied the parts sheet.

"I think I could take out another .7 of a kilogram," offered Bill—a pound and a half.

"What a guy," said Ron. "1.3 kilograms for me. Done. I have too many other things to fight over."

Someone asked if they really believed their numbers.

"I have a reasonable chance," said Bill.

"I . . . don't know," said Ron.

"Weight happens," said Doug. "Let's do the best we can to hit 79. We're not going to back off our schedule. I'd rather have a 2-pound failure than a two-week failure. But make like Richard Simmons and push it, baby."

The following Monday, March 6, Dean returned to Credit Suisse First Boston's offices in New York to sign the deal for Ginger over lunch in a small private dining room. There he found a thin, wary man with a hawkish face and white hair who looked like Andy Warhol with a better haircut. His name was Peter Huber, an engineer-turned-lawyer who also wrote conservative diatribes, most recently in a book called *Hard Green,* which argued that there were no environmental problems, only stupid or dishonest environmentalists. Dean rarely read nontechnical books, but admired Huber's and told him so.

A few minutes later, Jack Hennessy rushed into the room. He was the very model of a top executive: tall, tan, fit, handsome, with hair as white as his teeth. Trailing him was a portly man with a round face framed by bushy mutton-chop sideburns. Balding and double-chinned, he wore a dark purple suit with wide lavender stripes, a clashingly striped shirt, and a vivid bow tie. He looked like a cross between Mr. Pickwick and Austin Powers. His name was Myron Magnet. Presidential candidate George W. Bush had called Magnet's denunciation of 1960s liberalism, *The Dream and the Nightmare,* the most influential book he had ever read aside from the Bible. Magnet worked at the Manhattan Institute, a neoconservative think tank in New York, where Huber was a senior fellow. Hennessy was on the institute's board and was an important fund-raiser and advisor to the Bush campaign. Though Dean wasn't especially political, his views fit comfortably into the present company's. He once told me that he was a Republican at work and a Democrat at home, because Democrats invaded your business and Republicans invaded your private life.

Two large windows framed southern Manhattan and the World Trade Towers. Uniformed waiters brought in plates of salmon and shrimp salad. Between courses, Michael Schmertzler slipped in and sat down. Hennessy explained that he had invited this group to lunch so that he could pick their brains, especially Huber's, about an environmental strategy for Bush's campaign.

Dean, running on nervous energy, was intent on entertaining the table and began a long story about an idiotic lunch he attended at which Hillary Clinton wanted to pick the nonbrains of people like Barbra Streisand about health care. Hennessy and Magnet tried to interrupt several times, but Dean rolled over them. After fifteen minutes, when he launched into his spiel about FIRST, Hennessy barged in and asked Huber to position George W. on the environment so that Gore didn't beat him up on the issue.

That reminded Dean of the time he shared a taxi with Gore, that idiot, and off he rolled again, quipping that the plural of *anecdote* was not *data,* which somehow led him into education and, of course, back to FIRST. "And I happen to have some folders about it right here in my briefcase," he said.

Dean subsided a bit after that and let the political conversation swirl around him. At 2:00 Hennessy suddenly announced that he had to leave, surprising Dean. (Hennessy had to catch the Concorde for dinner in London.) Within minutes the only people in the room were Dean, Schmertzler, and I. Though Dean was here to sign the preliminary agreement for the Ginger money, he started to sell Schmertzler on the Stirling engine; he suspected that Johnson & Johnson was going to withdraw from the project. After listening politely for several minutes, Schmertzler reminded him that the business at hand was Ginger. He placed the papers and a pen on the table in front of Dean.

Dean was in no hurry to put a $30 million millstone on his back. He had some questions. He couldn't understand why CSFB could back out in forty-five days but he couldn't. He kept worrying the question, chewing on it, talking all around it. After fifteen minutes or so he finally said, "I don't *think* you'll back out—right?"

"No," said Schmertzler. "We want to do this. *Today,* so we can get started. We're excited about doing it *now.*" He glanced down at the agreement.

The two men couldn't have been more different. Schmertzler was crisp and direct, Dean prolix and roundabout. Schmertzler was cool and self-contained, Dean expressive and needful of assurances. Schmertzler's sentences were brief and premeditated. Dean's thought process was verbal and meandering.

Dean's next tangent concerned John Doerr. He told Schmertzler about meeting the venture capitalist at TED. Doerr wanted to see Ginger and had offered to send his jet to pick up a prototype. Harvard's Bill Sahlman had told Dean that Doerr never signed confidentiality agreements, to which Dean had replied, "Then there's no problem—he can't see Ginger." After talking at Schmertzler for a while, Dean eventually reached his destination: Would CSFB have any problem with Doerr investing as one of the individual angels, not as part of Kleiner Perkins? Schmertzler said no.

Which reminded Dean of Robert Halperin, who wanted in for $5 million and a board seat. So far, the board included Schmertzler, CEO and president Tim Adams, Bob Tuttle, Vern Loucks, and Paul Allaire. Alexander d'Arbeloff, former CEO and president of Teradyne Inc., an electronics and telecommunications firm, and now chairman of MIT's Board of Trustees, also wanted a seat.

Several of these men were older and semiretired, and Dean wondered whether Schmertzler thought it wise or unwise to fill the board with them. And what did he think of someone who seemed to make a board seat a condition of investment? Dean also wondered whether a couple of Sahlman's angels might be useful on the board—the guy from

Brazil's largest investment bank, Garantia, who knew South America, or the guy whose family ran three big industries in China.

Schmertzler tactfully opined that people with contacts in those parts of the world could be invaluable. Retired people, in Schmertzler's experience, were often troublesome because they used board seats as a hobby or to stay visible. Halperin and d'Arbeloff didn't yet know it, but they had just become history.

Dean asked about board members' compensation and other matters. Once or twice he seemed about to sign, riffling the agreement before veering off with another question. Schmertzler answered patiently, but exasperation occasionally flitted across his face. He kept saying, "This is your company. You will control it." He knew his client.

"It's possible we could lose the whole $30 million," Dean said, testing. "Not probable, but possible."

Schmertzler nodded with a slight smile. "Even if we did," he said, "which I don't think will happen either, it's $30 million out of a pool of $3 *billion* that we're working with."

Two hours after Schmertzler laid the agreement on the table, Dean abruptly took a pen from his pocket and pulled the papers to him. "It's not every day that I put on a $30 million pair of shackles," he said. As a last delay, he pushed the document over to Schmertzler. "*You* sign," he said. And then, with no room left for procrastination, Dean finally scrawled his signature. He looked agitated.

Schmertzler stood up. "Do you have somewhere you need to go?" he asked. "Because we've been holding a car for a couple of hours."

Dean looked surprised. "No, I'm in no rush." But Schmertzler was, and he stayed standing. Dean gathered his things. "You're the lion, and the lion will sleep with the lamb," said Dean. "But in the end, you're still the lion."

Two days later, another lion came prowling. John Doerr arrived in Manchester in one of Kleiner Perkins's big Citation X jets. At Dean's house he refused to sign the confidentiality agreement. Dean politely declined to show him Ginger and suggested dinner instead. Doerr's move: I want my lawyer to look over the agreement. Dean's countermove: Trust me, live on the edge.

Doerr surrendered, saw, rode, and surrendered again. He and Dean stayed up into the wee hours of what Dean later called "one of the most incredible nights of my life."

"And at the end," Dean told me the next day, "Doerr literally says—and he's got a good-sized ego, with good reason—'Dean, we've done more dot-coms than anyone. I never thought I'd see something in my life as big as the Internet, as far as making a difference. And I just saw it.'"

When a new idea interested Doerr, he went into hyper-mode. Over the next few days he besieged Dean with phone calls and e-mails, including this one with its in-a-rush spelling and exuberant adjectives about his first evening with Dean and Ginger: "It was amazing, mind-blowing, cosmic and therefore somehow extra-terrestial . . . but of course very down to earth. . . . [T]hanks for the scope of your ambition, which is breathtaking and inspirational."

Doerr made CSFB nervous. Jack Hennessy had told Dean that he could live with Doerr as an investor but wouldn't feel the same way about having his dominating presence on the board, just in case anyone was entertaining that possibility. The bank was supposed to take a month to do the final paperwork for the $30 million, but as soon as Hennessy heard that Doerr was stalking Ginger, he told Dean to expect the documents in ten days. Dean figured that Hennessy and Schmertzler were petrified that Doerr might try to poach the project.

They were right. Three days after his first visit, Doerr called Dean on a Sunday and asked if he could show Ginger to some Kleiner Perkins people the next day.

I'm busy, Dean said.

How about tomorrow night? asked Doerr.

No, can't.

What about Tuesday? pressed Doerr.

Sorry, I'll be gone until Tuesday night.

Until what time?

Around 8:00.

I'll be there at 8:00, said Doerr.

Tuesday evening two Kleiner Perkins Citation X jets arrived at Manchester Airport, the second one carrying KP partner Vinod Khosla, cofounder of Sun Microsystems and a VC legend in his own right. Once again they stayed up until 4:00 A.M., riding Ginger and imagining the future. Doerr's enthusiasm pumped up everyone. He called Dean a combination of Henry Ford and Thomas Edison, and promised to sponsor at least one regional competition for FIRST in Silicon Valley.

The question was how much of Ginger to give Doerr, who now wanted to bring in his partners at KP. CSFB wasn't going to reduce its share, which relegated Doerr to the $20 million pool with Sahlman's angels. Sahlman suggested giving Doerr and KP $7.5 million, by far the biggest private share. Doerr scoffed—that would make him a mere investor, and he wanted to be *involved*. He said he took pride in recognizing things that would cause discontinuities, and Ginger would cause a big one. He wished he could take the entire $50 million.

"I'm thinking to myself, *what* is going on," Dean said to me the day after Doerr's second visit. His head was spinning from the pace and turn of events. Less than three months ago he had been in the deepest funk of his life. Now people were begging him to take their money.

His only complaint was with the Ginger team's reaction when he told them about the CSFB deal. They had seemed relieved; Dean wanted them to feel added pressure. They had felt rich; Dean wanted them to feel obligated. They were happy that they could soon hire more people, buy needed equipment, not worry about every penny; Dean worried that the new cash would worsen what he considered their spendthrift ways. They had gone home early in celebration; he thought they should have worked even later than usual.

"Having someone put $50 million into your company doesn't make you rich," he said, "it gives you the *opportunity* to get rich." He had to keep teaching them that.

chapter seven

Tuck and Roll

March–April 2000

Mike Ferry, the director of marketing, had two assistants, Tobe Cohen and Matt Samson, a development that Doug Field referred to as "the khaki invasion." Tobe and Matt had been at Ginger for only a couple of months when they brought in two mechanical dolls and perched them on top of their adjoining cubicles. The dolls, named Tuck and Roll from the movie *A Bug's Life,* had infrared sensors in their bellies, which allowed them to talk and sing gibberish to each other in guttural, squawking tones. The engineers working across the aisle were amused, for a while.

Like the gibbering dolls, Tobe and Matt were talky and gregarious, constantly chatting from cubicle to cubicle, "ideating" and "capturing" bits of info. That's the way they worked. But engineers often need quiet to concentrate on a problem.

One day Tuck and Roll disappeared. Ransom notes began appearing on Tobe's and Matt's desks: "Cooperate or you'll never see your friends again." Then the demands began, for things such as a $7 gift certificate to a local Chinese restaurant: "Leave it in a plain envelope next to the Magna Doodle and nobody gets hurt." The marketers laughed at the notes and ignored the demands. So the next set of demands arrived with photographs of the dolls hooked up to a battery, or posed next to a blowtorch, or about to be crushed beneath a Ginger. Tobe and Matt laughed a little less confidently. Then the body parts started turning up—a head sliced cleanly off, an eviscerated torso. The accompanying note said, "You failed to meet the demands. You can still prevent the death of the other."

The mutilations upset Tobe and Matt. Engineers and marketers had such different temperaments. The marketers' meetings often resembled pep rallies, where the most common phrase was a hearty "Great idea!" Disagreements were couched in gentle language that preserved self-esteem: "I'm struggling with that" or "Can I lay out an alternative vision?" So different from the engineers, with their blunt, "*That* won't work" and "You want to do *what?*" A joke that featured disfigured dolls convulsed the engineers, whereas Tobe and Matt didn't understand how colleagues could inflict such aggressive humor.

Then one day Tuck and Roll mysteriously returned—the disemboweled doll had been a proxy—accompanied by a final note: "If you can't run with the big dogs, stay on the porch." The kidnappers remained unknown and at large, but Tuck and Roll never gibbered in the Ginger offices again.

Throughout this saga, the marketers had been busy ideating plans for Ginger. To impress Steven Spielberg, they had commissioned a video of professional skateboarders riding Gingers. One evening Tobe, Matt, Mike Ferry, Tim Adams, and Doug Field convened in Dean's home projection room for the premiere. The stunt riders took Ginger off jumps and talked about the experience in spacey vapidities: "It's cool," "It's fun," "It's weird." The scene switched to Easy Street, where several Ginger engineers pretended to be office workers or pedestrians. Despite the stunts and pumped-up music, the video looked amateurish and managed to make the machine boring. Even the lighting was bad.

When it ended there was dead silence. Dean bit his tongue. Show *this* to the director of *Jaws, Indiana Jones,* and *Jurassic Park*? "It was a *stupid* video," Dean said later, "for *$50,000*. A *waste* of money." Stupid and expensive: Dean's least favorite combination.

He was also frustrated that Mike Ferry had hired Lexicon, a professional naming company, instead of getting his team to devise a permanent name for Ginger. When Dean heard that Lexicon's services would cost about $70,000, he almost had a cow.

Ginger did need a serious name, something suitable for invoices and stationery. The previous June, before the marketers arrived, Dean, Tim, Doug, and Doug's small team had generated a long list of possible names, including such clunkers as Baltrans, Cogitrans, Edept, GRV, and Gyroporter. Doug wanted to stick with Ginger, but suggested other dancing names: Tango, Jitterbug, Arabesque, Cha-Cha. Dean didn't like any of these, but as a stopgap he had chosen a suggestion from his thesaurus, Donna Tamzarian: Acros. Lexicon was researching this vaguely classical word to make sure it wasn't offensive or ludicrous in any of the major languages, also known as customer bases. A visitor from Michelin had mentioned that the word was French slang for drug addict, but Dean and Tim Adams could live with that obscure blemish. They wanted to avoid big mistakes like the Chevy Nova, a source of hilarity in Spanish-speaking countries, where *no va* means "it doesn't go."

Mike Ferry had recently asked the team for more name suggestions and had gotten back duds such as E-Glide, Allgo, Weeble-Wobble, Bal-Pal, Ped-X, and Glideator. One engineer suggested using the names of ethnic dances from countries where the product might launch—in China, for instance, it could be Nishangyuyi, Siluhuaya, or Fantanpipa. Dean proposed Cyberwalker or the mathematical symbols for pi and torque, which gave the marketers a chance to grimace at him for a change.

Despite Dean's displeasure at the cost, Mike Ferry had decided that the team's suggestions made a compelling argument for hiring Lexicon.

In March the marketers took a research trip to California—a junket, in Dean's opinion, emphasis on junk. When they met with Dean a few days after returning, Dean stopped Mike Ferry's report almost immediately, right after Mike said they had tested a Zappy. Did the marketers really think that Ginger's competition was a little electric *scooter*? Why did it take three guys spending thousands of his dollars to eliminate Zappys as a competitor? That reminded Dean of the stupid $50,000 video. These guy were used to $20 million budgets that paid consultants to help sell paper towels. Maybe that was necessary when your quicker-picker-upper was just like fifty others, but these guys were selling something that nobody else had.

What would they do, he asked them, if they didn't have any money and instead had to rely on their most valuable asset—their own brains? Don't come to me with reports and data, he said. Use your imaginations. What if we gave away the first fifty Gingers, fancy ones with gold plate and Corinthian leather? One per day for fifty days, to high-profile people like President Clinton or the Pope, and to people who could open doors, or enhance our image on Wall Street, or soften the regulators. By the end of it, we would have gotten tons of publicity *and* created demand. Then we could auction another fifty for $25,000 each. That would help the bottom line, provide marketing information, pay your salaries, and throw Honda off the scent because Ginger would look like an expensive toy. If you're selling the world's coolest new technology, he asked, why introduce it like a new roll of toilet paper?

The marketers sat there silently. He knew they were thinking that radical technology couldn't gain public acceptance without a big push from a huge marketing budget. But the radical technologies that had taken a long time to find acceptance, like the PC and the cell phone, had been flawed and hard to use when they were introduced. Ginger wasn't like that. His elderly *mother* had learned how to ride in two minutes. And Ginger had simultaneous appeal—yuppies would want one and people needed it in India. And his marketers were talking about *Zappys*? (Ginger's engineers had bought and analyzed an extensive collection of such devices, but Dean approved of that because it was for engineering research.)

He figured he had pissed them off, so he took them to dinner to smooth things over. He kept talking, trying to shove them toward an approach that combined imagination and thrift instead of consultants and bloat. And damn if they didn't start taking notes. It was flattering but alarming. They should be telling *him* this stuff. God it was stifling sometimes, the way these guys thought about things.

Stifling did not describe John Doerr's way of thinking. Doerr had been stoking Dean's fantasies, telling him that in three to five years Ginger could be one of the biggest companies in the world, with profits of half a trillion dollars. Dean shook his head, telling me about it. "He's got me thinking too much about that," Dean said. "I have to stop it, because that's not the point."

But he liked Doerr's ambitions for Ginger because they matched his own—"so big-time compared to the bankers." If he and Doerr were right, the project was going to need an additional $50 million or so to launch. Doerr now wanted 5 percent of the company, a $25 million slice. Dean wanted to let him in at that level, but wasn't sure how Credit Suisse would react.

One day in March I walked into Dean's office just as he finished another call with Doerr. Dean was leaving later in the day for the FIRST regional in San Jose. Doerr wanted him to stay at his nearby compound. He also had offered to hold a cocktail party with some Silicon Valley hotshots whom Dean could hit up for FIRST. And since Kleiner Perkins now wanted a large percentage of the Stirling engine project in addition to a share of Ginger, Doerr proposed installing Vinod Khosla as Stirling's CEO for six months, to get the business up and running. "These guys are *out of control,*" Dean said happily.

He was in high spirits, perched in his office amidst the four pictures of Einstein and the chair painted to look like Einstein; the pictures of his airplanes and helicopters and private island; the life-size Darth Vader cutout leaning against the wall and the gigantic stuffed teddy bear lounging in a corner chair; the coffee table with machine parts on it; the big desk, always clean; the framed poster of the Boardwalk card from Monopoly; the poster next to it of a cartoon duck, wearing shades and sitting on a chaise longue with a drink, glancing at a fresh bullet hole just above his head.

But he was too charged up from Doerr's call to stay sitting. He jumped out of his chair to make his rounds. He spent much of each day wandering through his projects, asking questions, making suggestions, and just generally poking his nose in, like a bee spreading pollen, and sometimes pesticide, from flower to flower. Today he was headed to Ginger. He was going to be showing the machine to some KP people in California and wanted props—a chassis, a tire, a servomotor, the sketches of cops on the machine.

Doug, low-key and accommodating, saw to it. As they walked, Doug told Dean he wanted to hire a pair of specialists in electromagnetic interference for two days to analyze Ginger's design, at a cost of about $6,000. He didn't want to run into EMI problems later. Dean asked if they really needed the analysis. Yes, said Doug tersely. Couldn't someone at DEKA do it? asked Dean. No, said Doug, even more tersely.

"What exactly do they have to see?" asked Dean. Ah, that was his worry.

"The chassis and control shaft. I can strip off the wheels. They don't have to see Ginger."

"Do they work for the auto industry?"

"I don't think so."

"Can we trust them?"

"I think so."

Dean told Doug to find out how big the company was and who exactly would be coming. "And give them the riot-act speech before they sign the confidentiality agreement—that *any mention* of transportation is included. But don't get them excited." The engineers, like the marketers, often chafed at Dean's extreme strictures about secrecy.

We drove to the airport next. In preparation for Dean's trip to California, his jet was being serviced near New York City, and he had hired a small prop plane to fly him there for the pickup. As we waited in the private terminal, Dean filled me in on the latest intrigues with the investors. Jack Hennessy and Michael Schmertzler from Credit Suisse First Boston had finally agreed to let Doerr and Kleiner Perkins into Ginger. The bankers reasoned that if they did an initial public offering of the stock with Doerr's name attached, the opening price would double or triple. To me the most interesting part of that statement was Dean's "if." Hennessy and Schmertzler were worried, however, that Doerr might take over the deal. Schmertzler had called Bob Tuttle three times today asking about Dean's upcoming meeting with Doerr in California.

Dean was interrupted by his cell phone. Doerr again, wondering if it would be OK to bring one more person to their meeting—Steve Jobs.

"I only know him from folklore," said Dean, his eyebrows up. "He is certainly one of the visionaries, and also a maniac. Which in my opinion isn't necessarily bad. But do you want to bring an unquantifiable maniac inside the gates at this point? Do you know him?"

Doerr vouched for him and said he was also trying to get Jeff Bezos, founder and CEO of Amazon, to come on Thursday. Dean gave the OK and hung up. "Here's *John Doerr* chasing me around," he said in wonder. "Can you imagine a board with Doerr, Steve Jobs, and Jeff Bezos?" He cracked up. "Do you *believe* it? I keep thinking, 'Someone should write a book about this'—and then I remember that you are." He cracked up again.

After picking up his jet, he wanted a haircut before the California trip, so we went to his salon, Great Expectations. "Dean's here," said one of the stylists. His cutter wouldn't be free for ten minutes, so we walked over to a music store to pick up some new CDs for the plane.

He needed them. I had grown to hate Mary Chapin Carpenter's fine album *Stones in the Road* after hearing it repeat through the plane's earphones to and from wherever Dean happened to be going. He didn't seem to notice when the first song, "Why Walk When You Can Fly?" came around for the third or fourth time. In the helicopter, the repeating anthem was John Williams's music from *Star Wars*.

Today he bought Williams conducting music from Hollywood hits. I wondered which soundtrack would end up most appropriate for this trip to the land of dreams and venture capital—*The Wizard of Oz? Dances with Wolves? Jaws?*

Dean met Steve Jobs at Doerr's compound in the moneyed enclave of Atherton. Jobs didn't want to get off Ginger. He rode it hard all over Doerr's estate. Whenever Dean asked him to give it up for a while, Jobs would watch impatiently for a few minutes and then tell the rider, "Get off."

Dean thought Jobs was one of the most intense people he'd ever met. Jobs told him that Ginger was too big and important, that Dean would never control the idea for himself. The Chinese and Japanese would rip him off. If Dean really wanted to change the world and spare himself the misery of fighting pirates, Jobs advised, he should put Ginger into the public domain and charge a $100 royalty on every machine built. That way he would still make $50 billion, and in five years he would have changed the world.

No one mentioned investing. When Jobs left, Doerr walked out with him, then returned with the report: Jobs wanted in.

It's coming out of your piece, said Dean, not mine.

Jobs's forceful interest both exhilarated and worried Dean. The trip's highlight for him, however, was the FIRST regional in San Jose, where forty-three teams were competing. John Doerr got on stage there and promised to sponsor at least one regional competition next year. A few weeks later, Dean flew to Orlando for the FIRST championship at Epcot Center. Thanks largely to his relentless salesmanship and marketing, the organization had taken off.

Dean thought that one of America's biggest problems was its adulation of empty vessels like athletes and movie stars. Why celebrate people whose chief talent was dribbling a little ball or looking attractive in two dimensions? It offended him that kids knew the names of such people but not those of real heroes, the scientists and engineers who invented and perfected products that improved the world. Too many kids wasted their time and minds dreaming about playing in the NBA or starring in a music video. They might as well buy a lottery ticket.

Meanwhile, thousands of engineering and technology jobs went unfilled every year. Many of the brightest students were choosing law school or business school, but lawyers and managers and stockbrokers didn't create products, and products were what created wealth. The engine that drove the country's growth and greatness had always been technology. But now American students scored abysmally in science and math compared to their peers in other countries. Dean worried that scientific illiteracy would wreck the country's economy, lifestyle, and future. He wanted to wean kids away from the artificial sustenance of celebrity and onto the protein of engineering and technology.

So in 1992 he convinced several companies to pair their engineers with high school students to build robots for a competition he called FIRST. That year, twenty-eight teams competed in a New Hampshire high school gym. Dean told anyone who would listen that he intended not just to make the FIRST championship bigger than the Super Bowl, but to change the culture of the United States. He meant it. He pursued corporate funders relentlessly and shamelessly. More than three hundred major companies now sponsored teams, including eight of the Fortune 500's top ten.

When Dean flew to Orlando in early April 2000, more than 20,000 kids on 372 teams from 41 states were participating, and the championship was Epcot's biggest annual event. Epcot built FIRST a special stadium that held thousands of spectators. In some schools, it had become tougher to make the FIRST team than the football team. Schools sent off their FIRSTers with pep rallies and hung banners when they won. Engineering colleges had started coming to FIRST competitions to recruit students and offer scholarships. Nevertheless, Dean was a little disappointed. It was taking longer to change the culture of America than he had expected.

The championship at Epcot resembled a Roman circus for science students. There was frenzied cheering, loud music, wild dancing, flying Frisbees, packs of kids roaming around in team T-shirts and bizarre costumes. The atmosphere mixed earnestness and revelry, intense competition and delirium. Matches occurred simultaneously on several courts. Triumphant teams pumped their fists and hugged as cheerleaders shook pom-poms and the vanquished silently pushed their battered robots to the side. The pit area was a hive where teachers, engineers, and students plotted strategy for their next match, or frantically repaired their machines. The place buzzed with energy, like the air beneath power lines. And the stimulus of all that youthful energy was science and engineering.

Dean ate it up like ice cream. FIRST reflected his best traits. "Just another one of my crazy ideas that got a little out of hand," he said, laughing. He was a superstar here, mobbed wherever he went, mostly by kids but also by teachers and engineers. They wanted to talk to him, to thank him for changing their lives, to get his autograph on their T-shirts, or just to stare at him with moony eyes. He responded with easygoing charm, asking questions and making jokes. But he never forgot to be a Pied Piper for engineering, always urging them to demand more teams from their school districts.

There were astonishing stories here. For instance, the team from Broadway High School in San Jose. Broadway was the last stop on the way down for four hundred kids who had been deemed failures and disrupters at other schools. A counselor from Broadway told me that if NASA hadn't started sponsoring a FIRST team there last year, not one kid here today would have graduated. "These are life-altering experiences," she said.

Steven Longo's life certainly had been altered since joining the team the previous year. His porkpie hat, pulled low over his watchful eyes, was no wannabe's costume. A gang member from age eight, he had been stabbed twice in the head and twice in the body. He plummeted into Broadway with an academic average of 0.1. Then FIRST arrived. Longo liked to take things apart, so he checked out the program. It snared him. Team members needed to maintain a B average and good attendance. So he did. He loved the challenge of building a robot and became adept at wiring and programming. Like the rest of the team, he often worked on the machine until 10:00 at night.

He showed up at the regional last year with bandaged arms—he had gotten more than twenty gang tattoos burned off because he didn't want to embarrass the mentoring engineers from NASA. He had dropped out of the gang. Now a senior, he had a 3.8 average and intended to get a master's in mechanical engineering. "I have a new life now," he told me, "a different life." There were similar stories all over Epcot.

At the fireworks display after the competition, I ran into Michael Schmertzler from Credit Suisse. He was always cordial but clearly would have preferred me to disappear from the project. I joked that I wasn't sure of his first name because Dean, as a mark of his respect, always called him Mr. Schmertzler. Dean overheard me.

"That's only until I pay him back," said Dean. "Then I'll call him Mike and he'll call me Mr. Kamen."

While Dean promoted Ginger in Silicon Valley and FIRST in Florida, Ginger's marketing team had been working on the "homework assignment" he had given them: Make a list of prominent people who might merit a free Ginger. They found the idea brilliant, but worried about positioning the machine as a toy for rich people. They also wondered how Dean expected to score a major newspaper article or TV story every day for two months, something they had never seen happen. They didn't think Dean understood marketing, either in the abstract or the particular, though his ideas and instincts could be amazing. Yet they considered Dean their biggest asset and greatest salesman. "The number of Gingers we sell," said Tobe Cohen, "depends on how hard Dean wants to work at it."

When they met with Dean about the homework assignment, they finally got to give him the report about California that he had aborted before. Yes, they had rented Zappys, but also two electric minicars, DaimlerChrysler's GEM and Nissan's EV. Bystanders were very curious about the cars, which suggested a receptive market for alternative vehicles.

"We gotta keep telling ourselves we are *not* going into the street," said Dean, hitting the table for emphasis. "We are an *electric pedestrian.* It's *not* a vehicle. It's an alternative to walking. We *can't* let them call us a scooter." That's why they had to give Gingers to important people who, when told by a cop to get off the sidewalk, would go straight to the mayor or the police commissioner.

Dean also had an idea about connecting Ginger with the Disney empire. About 45 million people a year visited Disney's theme parks. These were people who could afford to get on a plane, stay in a hotel, and spend $40 per kid per day on rides. Dean had learned from someone at Disney that the theme parks now rented three thousand wheelchairs every day at $25 per day, and 80 percent of the renters could walk but preferred not to. Many of those people would love to rent Gingers. Disney would generate great publicity for the machine. Maybe the company would even sell Gingers in their eleven hundred stores across the country.

The marketers looked gleeful, but Dean wasn't finished. He had connections at Disney, of course, because the company loved FIRST. And Disney had released three huge hits from Pixar Studios, *Toy Story, A Bug's Life,* and *Toy Story 2.* Pixar happened to be owned by Steve Jobs, freshly amazed by Ginger. Dean had suggested to Jobs that Ginger might look great in a Pixar movie someday. And John Doerr had offered to get Ginger's Web site done by a top team.

The marketers were agog. Dean's only bad news was that the filming of *Minority Report* had been postponed while Steven Spielberg recuperated from getting a kidney removed.

The marketers began presenting their list of famous prospects. Dean drew a blank on most of the names from entertainment: Nicole Kidman, Martha Stewart, Regis Philbin, Brad Pitt, Rosie O'Donnell, Tom Hanks. He asked if Madonna was a first name or a last name. He nixed rock singers and pushed for heads of state, environmentalists, business leaders. How about the president of New York City's taxi and limo union? On second thought, probably not. "Because I think they'll spend more than God to crush us," he said. "I think we're the cab system's worst nightmare."

The meeting had started at 5:00. It broke up at 10:00. The marketers, who had learned their lesson from previous occasions, had come stocked with chips and candy bars. As usual after meeting with Dean, they left feeling battered, inspired, and frustrated.

At dinner afterwards Dean fretted about them. He worried what would happen when Doerr met them and Tim Adams. "I'm afraid he'll think that Tim is from the Rust Belt and the marketing guys are from old manufacturing." He was already starting to look at them through Doerr's eyes.

It was 11:30 by the time we reached his hilltop house. Near it, Dean had built New Hampshire's largest wind turbine. In a 24-knot wind it produced 84 kilowatts of power. Its electricity helped run the house and sometimes produced a surplus, which Dean enjoyed selling to the utility company. On this windy April night, the turbine's blades were whirling. He looked up. "I'm making power," he said with a big grin.

chapter eight

Deaned

April–June 1, 2000

While Dean was at Epcot transforming the culture of America, Ginger's engineers were up to their elbows in more tangible labor. Thickets of problems stood between them and the May 31 deadline for D1. So many pieces of the machine had to change, and every change cascaded through the rest of the design. Conversations began with "Can you," "Can we," or "What if," and ran smack into "But then," "Except that," or a simple "No way." But there had to be a way, and they had to find it fast. They had promised Doug and themselves. Besides, engineers believe there's always a way.

Phil LeMay, the lead engineer for Ginger's electrical systems and sensors, was worried about the FETs, or field effect transistors—the amplifiers that power the motors. A FET looks like a mechanical beetle: a thin black rectangle, about three-quarters of an inch long and half an inch wide, with curved metal prongs sticking from one of the narrow ends. Every time Ginger accelerated, the FETs handled jolts of high current, which also created heat. In the existing design, the twenty-four FETs hung from the circuit board. If they somehow got bent and touched metal, they would short out and abruptly kill the motors. The rider, however, would keep right on going, over the handlebar.

Phil wanted to change the design, but two "bosses"—hollow cylinders in the chassis that received screws—blocked his path. He stared at the bosses, picking at them like bothersome warts. If he could move the FETs near the outer edge of the boards, the power would flow more neatly, the heat would dissipate more quickly, and there would be less electronic noise to bother the neighbors. Phil called it "bad city planning" to run the current through the middle of everything else, like putting a highway through downtown—it was noisy, dirty, and unsightly. Bad city planning also caused bottlenecks, dangerous intersections, bickering neighbors, accidents. Phil kept picking at those bosses, damn them.

A few days later he handed the problem to J. D. Heinzmann. On this April day, J. D. was in the Penthouse lab, peering at an electronics board secured to a table. Nose clamps fastened the board to several testing devices, like a patient in surgery. J. D. had started redesigning the board, a complex undertaking, like platting a new city with streets, traffic controls, speedways, quiet neighborhoods, and industrial zones. Engineers described finished board-designs as "fully populated." In J. D.'s first whack at it, the FETs and diodes were surrounded by subdivisions of tiny silver resistors, capacitors, and transistors. The connecting red wires were 1/100th of an inch thick. One oversized FET, the queen bee that controlled the current from the battery, was nicknamed the BAF: Big Ass FET.

Above J. D., two analog wall clocks labeled Manchester and Ulaangom, Mongolia, showed exactly the same time. On the wall, a movie poster advertised Fred Astaire and Ginger Rogers waltzing on roller skates in *Shall We Dance*. Some of Ginger's alleged competitors huddled nearby: the EV Rider, the Go-Ped, the Razor, the Zip, the Xootr, the City Bug, and Honda's Raccoon and Gyro (a three-wheeled scooter). People test-rode Gingers in a room that opened off of the Penthouse lab. The walls were scarred with holes and black tire streaks made by errant Gingers. Doug Field asked people to sign and date the bigger blemishes. My embarrassing perforation said "Steve K, 12-6-99."

For several weeks, the path of power had been giving J. D. fits. Each Ginger would have two motor-drive boards, one for each of the independent electric servomotors. The motors, about the size of a small can of malt liquor, were capable of 2 horsepower each, making them the most powerful in the world for their size. J. D. described the motor drivers as the machine's electric muscles, the things that made it go. He was stumbling along, banging his head a lot, but every time he redesigned the boards he learned so much that he stayed jazzed. He worked late and got up early, routinely logging sixteen-hour days, but he didn't feel tired. "I essentially have no life outside DEKA right now," he said. "It's hard to remember that other people do have lives." During this period he won Dean's monthly award for most hours worked in one week—106. Doug Field was next with 94.

When J. D. got stuck he snagged a few hours of help from an electronics whiz at DEKA. If neither of them could figure it out, J. D. called in the expert from the company making the motors. The expert knew motor controllers the way Clausewitz knew war. He would look at J. D.'s newest board with a sardonic smile and mutter, "Kluge," then force him to redesign it using math instead of solder and instinct. Engineers just want to make something work, but this expert pushed J. D. and the DEKA whiz to *understand*. These sessions invigorated and humbled them.

"I felt like a fish out of water yesterday," said J. D. in the cafeteria after one such session.

"Me too," said the DEKA whiz.

"But you understood what he was talking about."

"No I didn't."

"You mean you were faking it too?"

Sometimes J. D. mistakenly assumed the fault was his, as with the chips. Every chip had eight pins, four per side. Each pin did a different job, which was labeled and described on a diagram called a pin-out. J. D. had been getting baffling data that drove him toward decisions that led to more baffling data or to burned-out chips. After three frustrating months, he discovered the problem—the manufacturer's pin-out was wrong for three of the chip's pins.

One day in April, at lunch, he explained that he was trying to track down the cause of a delay when the boards commanded the motors to make an instantaneous current change, which translated into an instantaneous torque change at the wheel—in other words, the moment when a rider asked the machine to accelerate or decelerate sharply. The delay was 200 microseconds, about 2/10,000ths of a second. "That's a *lot* of time," said J. D., sounding affronted.

Dean swooped into the cafeteria. He had heard about a new type of carbon battery and wanted J. D. to evaluate the specs, make some calls, and report to him tomorrow morning. Then he accelerated off.

"I just got Deaned," said J. D. Getting Deaned was the DEKA equivalent of an instantaneous torque change.

Next Dean swooped into the Ginger area and began haranguing Doug Field and industrial designers Scott Waters and Tao Chang. Dean had been thinking about bureaucrats. If Ginger couldn't get on the sidewalks, it was dead. So he had a few design ideas. How about a soft Ginger? Cover all the steel with foam, shroud the wheels almost to the ground with foam fenders, and wrap a cushy control shaft around the rider's waist. That way if Ginger hit a pedestrian, it wouldn't hurt. "Can you give me a couple of nice pictures of something like that?" said Dean. Then he accelerated off.

Scott and his assistant Tao were already up against the wall, working many dozens of hours a week on D1. Now Dean had suggested a design that was not only completely different, but aesthetically hideous. They had just been Deaned. Tao was new and looked worried.

Doug asked them to step into the War Room and closed the door. "This will give me a chance to explain to Tao what just happened," said Doug, smiling. "You were just sucked into the Dean vortex. It happens and it will happen again. Dean's success comes from the fact that he's completely nonlinear. He'll get off the highway and explore the side roads before getting back on the highway. So don't be alarmed. Dean jokes about it. He says, 'I tell people to do things ten times a day, any one of which, if it actually got done, could sink the business.' "

The way Doug read it, Dean wanted an alternative to show investors and regulators in case it began to look as if Ginger might get pushed off the sidewalks. Scott fretted about the time this exercise would take away from D1. Doug chuckled that of course Dean didn't expect anything else to slip because of his request. Doug suggested that they sketch a couple of rough concepts, nothing too time-consuming, then sit down with Dean over pizza. If he liked anything, they could do more. "This is pure frog-kissing," he said, still smiling. "Use it as a fun diversion, not a distraction." He asked them to do it in the next few days, before Dean asked again. Oh, and one more thing—don't mention Dean's design idea to anyone else on the team. Doug smiled again. "There's no point in freaking out people who are already staying to midnight."

In his characteristic style of leadership, Doug had reacted quickly and reassuringly, as if Dean's request was droll rather than traumatic. He defused their panic with explanations, gave them clear instructions, and acted confident that this trifle wouldn't pressurize their real schedule. He turned Dean's bomb into a firecracker. Scott and Tao left the War Room feeling relieved instead of mangled.

Dean supplied Doug with plenty of opportunities to practice such management tactics. A few days after Dean's suggestion for a "soft Ginger" (soon nicknamed the Nerf Ginger), Doug relayed one of Dean's redesign ideas about the chassis to Ron Reich, who went rigid. Doug put his hands on Ron's shoulders and smiled. "Don't worry about it," he said, "I just told you so you'd appreciate my insulating properties." No wonder the team would do anything for Doug.

If Dean was the project's patriarch, Doug was its conscientious big brother. Or, as he liked to describe himself, its soul-keeper. He sometimes said his role was to convince the entire team to believe that they were on a mission from God. He stayed alert for anything that threatened his vision for the machine.

Vision, soul, culture, mission—Doug had thought hard about all these words as they applied to Ginger, and he spoke them earnestly.

"I want this to be a story I tell my grandkids about," he said. "I feel *tremendous* responsibility to the technology and the product and the people here. That can be overwhelming at times." Though he had a pregnant wife and two small children, he was spending at least twelve hours a day at Ginger and another twelve over the weekend. "But I don't ever stop thinking about it," he said. "It's an obsession."

Every day there were countless conversations among the engineers about countless details. Every centimeter of the machine was debated, fussed over, cursed. Once, some engineers were talking about how to attach a small metal tab to the lower control shaft.

ENGINEER #1: The alignment of that weld has to be within 50/1000ths. Can that be done reliably and repeatedly?

ENGINEER #2: This thing is a nightmare. We're adding something that won't allow tolerances.

ENGINEER #3: Can we add a visual? [To make the alignment more accurate.]

#1: Even if the visual is a big red stripe, it's going to be 1/10th of an inch correct or incorrect. A visual is not deterministic. A *bolt* is deterministic.

#3: I'm worried about rejection rates.

#2: It's going to be hard to find a place that can do it to these tolerances.

#1: It's an interesting tolerance problem.

#2: Interesting as in it hurts. This is too hard. I quit. You said it would be easy to build a balancing machine.

#1: It is. It's putting on a tab that's hard.

Though the pressure had notched up because of the D1 deadline, the team rarely blew a gasket. Disagreements were sometimes strong but rarely heated. The two who butted heads most often were Ron Reich, the lead mechanical engineer, and Scott Waters, the industrial designer. Their personalities abraded each other, and their different missions added to the friction. Ron's goal was to make sure that every part in Ginger was cheap, readily available, and easily assembled with the others. Scott's was to make every part beautiful and to create a harmonious whole. They spoke different languages—mass versus visual weight, electromechanical noise versus design noise. Ron was in his early forties and more experienced than the twenty-something Scott, and he sometimes patronized the young designer. In the first months of the project, many of their disagreements occurred because Ron made a unilateral irksome decision that Scott heard about later.

Doug Field recognized the problem from his days at Fred. "These guys are all so different in outlook, in personality, in ways of working," he said, "and they have different ways of talking to each other. The standard engineer's response to conflict is what I call 'the retreat to independence': 'I can't deal with these people, so I'm going to go off and just *do* it.' The rugged individualist versus the community."

Doug respected individualism—like Dean, he hired the most high-powered people he could find—but he insisted on community and communication. He warned Ron and Scott that if they ruined their relationship, he would be able to point to places in the design that had suffered because of it. He asked them to put the product and the team above their disagreements.

To nudge them in that direction, he rearranged Ginger's floor plan and seated them next to each other. Scott resisted, but Doug sold him on becoming a design evangelist among the heathen engineers. More practically, Doug wanted Scott and Ron to overhear each other's conversations and catch irritants before they got inflamed. If Scott heard Ron tell someone, "Here's where we'll put the parting line" (the welt where a die's two halves meet, or the seam left from a weld), Scott could immediately say, "*There*? No!"

The plan worked. Ron began consulting Scott about his tooling decisions. Scott checked with Ron about the cost and manufacturability of his design ideas. The effects rippled outward. Ron modulated his horsepower in meetings and curbed his tendency to run over people. "In Detroit there were times I could bully my way through, pushing and pushing and pushing," he said. "Here I can't operate that way. I have to be able to explain *why*. I think I've changed. I think Doug's happy."

One day near the deadline for D1, Ron, Scott, and Doug discussed a tweak to the chassis design that Scott wanted. When Doug objected to the cost, Ron jumped in on Scott's side. Doug's face reddened with suppressed laughter. "This conversation is surreal," he said. "I'm in an unaccustomed position." Ron, catching on, said, "I'm working for you, Scott," and smiled at Doug: "Isn't this what you've wanted?" It was. The team was better for it and so was Ginger, though Ron and Scott never stopped butting heads.

"The way you put people together has a huge impact on what you get in the end," Doug said. "Great design doesn't come from writing specs. It comes from a *culture,* from good people who understand that it's multidisciplinary. But it's hard to operate that way, because there's bound to be lots of conflict. There should be. My job is to deal with it early."

New Agers speak of their "spirit animal," the creature that reflects their inner selves. Doug, an engineer to the core, thought of his team members in terms of their inner cars. Ron Reich, loud and abrasive but with tremendous raw horsepower, was a Rausch Mustang with glass-pack mufflers. He disturbed the neighbors, and if he was speeding on a curvy road, he might end up in a ditch. But if the road was fairly straight and the destination clear, nobody got there faster.

J. D. Heinzmann, on the other hand, resembled one of the cars he owned, an old Mini Cooper—noncomformist, unfrilly, but well engineered, a versatile economy car that could be turned into a racing machine. Mike Martin, the quiet engineer in charge of structural integrity, reminded Doug of a Honda—reliable and thoughtfully built. John Morrell, the controls engineer, was a Ferrari, high-performance and high-maintenance, emotional and passionate. Bill Arling, whom Doug described as the team's "retro grouch," was a Porsche 911, which engineers considered a triumph of development over design. Like the Porsche, Bill would stick to nonconformity for its own sake, even if there were an easier way, and still make it work.

I asked Doug to name Dean's inner car. He laughed and shook his head. "Can't do it. He's *so* complicated." And his own? He asked me to suggest something. He drove a neon green Volkswagen GTI that was always at Ginger no matter how early in the morning I arrived. He liked it because it was high-performance without being showy. I said he seemed more like a small Mercedes or Lexus SUV—elegant but adaptable to rough terrain, and able to get wherever it needed to go. He liked that.

It was mid-April. The plan laid out more than a year ago assumed that by this point Doug would have twenty-eight engineers. He had fourteen. Now, six weeks before the D1 deadline of May 31, Doug saw signs he didn't like. The team seemed overwhelmed by too much to do in too short a time. This was compounded by a couple of suppliers whose parts were late. The combination of stress and outside delays caused some lassitude. If other people seem likely to miss their dates, why kill myself to meet mine? Doug thought of this as a disease that slowly paralyzed a team. He had seen it happen at Ford and at Fred.

But if the team missed the May 31 deadline because it really was impossible, I asked, wouldn't that hurt morale too? Doug looked fierce. "That is *not* a possibility," he said. "We'll do whatever it takes. An event like this is a ratchet—you close it behind you and move forward."

He knew that some people would be pulling all-nighters to meet the deadline for D1. Fine. When the next big deadline loomed, they would resist the engineering tendency to tinker. The D1 deadline, he told the team, was a test of their credibility not only within DEKA but within itself. Success would breed confidence. The surest way to avoid being micromanaged by Dean was to hit their dates. They *would* make the deadline, Doug insisted. "The only question is how much blood we'll lose making it."

Doug, Scott Waters, and Phil LeMay were in the War Room talking about Ginger's user interface, essentially the dashboard on the handlebar. The design called for a small module containing an electronics board to fit into the hole on top of the control shaft. That would allow the connecting wires to run down the shaft to the machine's control center in the chassis. For aesthetics, Scott wanted the dashboard module to be elliptical. That meant that the top of the shaft had to be elliptical, which presented problems of casting, sealing, and connectivity. Engineers prefer right angles and don't mind circles, but ellipticals are one of those industrial design notions that make them shudder.

They began sketching possibilities. "Maybe if the board wasn't flat," Phil muttered to himself, "if I put it into a flexed circuit . . ." Doug and Scott perked up. "That changes *everything*," said Doug. "A curved board," said Scott, getting excited. Design horizons suddenly opened up on this 4-inch space.

They covered drafting sheets with ideas, reaching in to draw on each other's sketches. A flexed circuit would allow for better sealing, a safer weep hole for drainage, and fewer parts. They always pursued the same goals: simpler, fewer, cheaper, easier.

Spring had finally reached New Hampshire, and the War Room's windows were open for the first time in months. A fresh breeze nuzzled the sketches on the table. Doug's Magna Doodle said "42 Days to D1."

Ron Reich handed Doug the new prototype of the chassis. "It's 95 percent right," said Ron, a high compliment from him. Doug sniffed the part. "I can still smell it, it's so fresh," he said.

Engineers love parts the way chefs love ingredients. They ogle parts, fondle them, stroke them, tap them, pass them around, put their ears to them when they move, and even smell them when they're fresh.

The new chassis prototype had come straight from DEKA's Stereo-Lithography machine. This machine used computerized lasers to turn engineering drawings into three-dimensional parts made of hard resin. The process, known as rapid prototyping, had transformed engineering design. It was fast and inexpensive. It shortened the design process, because engineers could examine fabricated parts within hours instead of days, detect mistakes and flaws, and make quick changes, all before committing big bucks to molds and dies. The tooling for Ginger would be pricey, so the Ginger team had to be sure the parts connected properly.

"Let's go see if the gyros fit," said Doug. The gyro assembly needed to sit in a small cavity in the center of the chassis. The engineers called this assembly the HOG, short for House of Gyros. Phil LeMay and his team had been working hard to shrink the HOG to the space available for it. A few days earlier, Phil had stared at a computer drawing of the chassis for several minutes and then asked the design engineer why there seemed to be unclaimed space along the HOG's four walls. "You can have it," said the engineer.

"I'll *take* it!" said Phil, excited. His bonanza was 2 millimeters in each of the HOG's corners—8/100ths of an inch.

"Drum roll, please," said Ron as Doug held the cubic HOG over the chassis. When Doug let go, it dropped snugly into its hole. Ron leaned over for a look. "*Loads* of clearance!" he said. "A couple of 50ths." That's 50ths of an inch, the edge of a credit card. Phil's bonanza had paid off.

"There's nothing like the satisfaction of seeing two parts come together," said Doug. Within moments, seven engineers had gathered around to enjoy the sight.

One evening near the end of April, I walked into Dean's office and found him more energized than usual because of three phone calls that day. Steven Spielberg's office had called again, and Dean was going to California next week to meet the director. Spielberg expected to start filming *Minority Report* in July and release it in the summer of 2001. That would coincide with the launch of Ginger, the launch of the iBOT, and a movie that actor Noah Wyle planned to do about Dean and FIRST. "We are going to *dominate* the world next summer," said Dean.

Shortly after he heard from Spielberg's office, John Doerr rang. Doerr wanted a bigger share of Ginger. Jack Hennessy and Michael Schmertzler at Credit Suisse First Boston had agreed to let Doerr and Kleiner Perkins in for $25 million, but wanted to maintain CSFB's status as the dominant investor by at least 30 percent over KP, which meant upping the bank's share of Ginger from $30 million to $39 million. Dean didn't want to give it to them, and said so to Doerr. "Give it to *me!*" Doerr replied.

The reason he had called Dean, in fact, was to push for an *additional* $25 million. He also wanted a controlling interest in the Stirling engine project, and now wanted to install a veteran of several Silicon Valley start-ups as Stirling's CEO. He added that Steve Jobs couldn't stop talking about Ginger and wanted to see Dean in California to discuss investing. Dean didn't know what he was going to do about any of this, but damn, it was exciting.

He went to California the following week. I found him in his office the day after he returned. He was still spinning. The Spielberg meeting had been a bust. The director didn't show up, but his people tried to pressure Dean into showing them Ginger anyway. He walked out. Spielberg knew where to find him. (*Minority Report* finally appeared in the summer of 2002, with no bit parts filled by the iBOT or Ginger.)

But there were other dreams to chase in California. Last night he had eaten dinner at Steve Jobs's house. Like Doerr, Jobs told Dean he shouldn't be letting a Rust Belter like Tim Adams (whom Jobs had never met) run Ginger. He also sneered at Ginger's design. But he thought the machine was as original and enthralling as the PC, and felt he *had* to be involved, which for him meant owning a significant piece of the company. He offered $25 million, evidently the mark of significance among West Coast investors. Dean hedged, suggesting that Jobs convince Doerr to give him part of KP's share.

But the main reason he was still spinning was that just a few minutes earlier, Doerr had called again to say that Jobs now wanted to put in $50 million. Dean showed me the little yellow Post-It where he had written down Doerr's latest figures and percentages: CSFB, $30 million (6 percent); Sahlman's angels, reduced from $20 million to $15 million (3 percent); KP, lifted to $40 million (8 percent); Jobs, $50 million (10 percent). The total came to $135 million and 27 percent of the company.

"That ain't gonna happen," said Dean. He had told Doerr the same thing. You're crazy, Doerr had replied—anyone else would jump at $50 million from Steve Jobs. Nor was Dean willing to give KP $40 million. Michael Schmertzler would have an infarction if he even suggested it. Schmertzler, in fact, was down the hall today with Bob Tuttle, working on the CSFB paperwork.

A few days later, on Friday, Doerr called again to lobby Dean about Jobs. He used an arsenal of tactics. First, flattery: Jobs has never taken an interest in anyone else's company until this and, like you, he's a visionary who wants to change the world. Next, celebrity: Jobs is a legend and you should be grateful for his attention. And finally, threat: If you aren't interested in involving Jobs, KP might not be interested in you.

That night Doerr sent a fax with his revised suggestions: KP had jumped to $50 million, Jobs to $63 million, CSFB to $38 million. Sahlman's angels kept shrinking, to $12.5 million. Doerr wrote that Jobs would sweeten things for Dean and CSFB by buying in at a considerably higher price per share, taking only 10 percent of the company for his $63 million.

The figures were so out of line with what had been discussed that Dean wondered if this was Doerr's way of backing out. Dot-com stocks had collapsed a few months before. KP's portfolio must have been devastated. Maybe Doerr and his partners couldn't afford Ginger anymore.

Dean couldn't sleep that night. The next morning, Saturday, he called Doerr at home. "I figured I was playing for all the marbles," he said later. He asked Doerr to answer a simple question: If Jobs wasn't included, did KP want out? Doerr quickly said no, but added that Jobs was a wild man with no throttle, and wouldn't be involved if he wasn't the main investor. Doerr urged Dean to make room for Jobs by dumping CSFB. Dean refused. He wouldn't break his word to Hennessy and Schmertzler.

Besides, Dean was torn. He sensed that he and Jobs resembled each other in some ways. Both of them were less interested in the money than in the vision. Jobs would make a world-class partner, bringing glamour, high-tech credibility, and intense genius to the project. But letting a wild man with no throttle into Ginger made Dean nervous. "This is a guy famous for firing people in elevators," he said.

Doerr called back that Saturday afternoon to press the case for Jobs. As they talked, Jobs beeped Doerr. What a coincidence. Jobs joined the conversation and the two of them worked on Dean from both flanks. Dean asked why Jobs needed 10 percent. Because I can't stop thinking about Ginger, said Jobs. He either had to get deeply involved or tear himself away. And since he expected to be more involved than anyone else, he needed at least 10 percent for his time and energy. Doerr magnanimously offered to help with Jobs's piece by cutting KP's share from $50 million to $30 million. This sacrifice didn't impress Dean, since he had only offered KP $25 million in the first place.

When Dean held his ground, Jobs made a graceful retreat. Maybe he could live with 5 percent, he said. Maybe they all should rethink things and talk again in a few days. Dean hung up with relief. But he didn't know what he would say next time, no matter what percentage Jobs wanted. And he had no idea how to talk about any of it to Michael Schmertzler, who would detonate if he knew about Doerr's fax.

"This is bizarre," said Dean, handing Doerr's fax to me. We were in his library late one night, a few days after the weekend phone calls. "The market is way down. There are no IPOs. Start-ups are happy if they get one or two million dollars, and companies looking for a second round of financing are dying on the vine." He shook his head. "And here we are with a higher valuation and a higher price per share than two months ago."

As the dot-com collapse made clear, there was a lot of abracadabra in the valuation of a start-up. But CSFB and KP both agreed with Dean and Bob Tuttle's valuation of Ginger: $500 million. It was the largest valuation either company had ever given to a start-up. And now, using Doerr's figures, it had jumped to $630 million. Dean was getting theoretically richer every day. "But that's *not* what it's about for me," he quickly added.

It was late, but he called Bob Tuttle for another discussion about Doerr, Jobs, and Schmertzler. Dean mostly listened, occasionally asking questions: "What did he say then?" "What should I say to that?" It was odd to hear him sound reliant. By the end of the conversation the two of them had devised a new split. Kleiner Perkins, Credit Suisse, and Jobs each would put up $38 million for 5 percent of the company.

Sahlman's angels would put up $12.5 million for roughly 2 percent. That came to $126.5 million and about 17 percent of Ginger. Now all they had to do was sell the idea to the investors.

A week and a half after the conference call between Dean, Doerr, and Jobs, Jobs phoned Dean at midnight. He hadn't called earlier, he said, because Apple's stock price had dropped more than 10 percent in the past week, so he had been busy with that and Pixar and his family. He said he really didn't have time to be involved with Ginger. Dean felt both relieved and disappointed.

But, continued Jobs, his passion and instinct wouldn't let Ginger go. He found himself thinking about it even in the midst of everything else. Dean thought he wanted in again. But, continued Jobs, you're uncomfortable giving up a lot of the business, which is understandable. It sounded to Dean that he was backing away once more.

Then Jobs made a remarkable offer. He volunteered to be deeply involved as an unofficial advisor for six months. At the end of that time, they could make a deal or not, depending on Dean's comfort level with him. Dean hung up feeling elated. He would be getting Jobs's insight and counsel for free.

Michael Schmertzler of CSFB had been alarmed when Dean told him about Jobs's offer of $63 million, but Dean assured him that he would never accept it. Dean relished how different this deal continued to be. Most start-ups didn't have much leverage with investors, but in this case all the potential investors knew that if one of them tried something shady or threatened to drop out, Dean would simply replace them without missing a beat. Schmertzler and Doerr called constantly to make sure they weren't losing their places in line. Dean had heard lots of stories about how Doerr was such a mystical guy, unreachable by mere mortals, but he was calling Dean several times a day from all over the globe.

After Jobs's call, Dean and Bob Tuttle put together yet another plan. Dean went over the figures with me one night in his library. Ginger would have 100 million shares worth $5 per share, giving the company a valuation of $500 million. Credit Suisse and Kleiner Perkins each would get 6 million shares for $30 million. The angels would take 3 million shares for $15 million. Dean reserved 5 million shares for future distribution. Ginger's employees would split up another 2 million shares. All 177 people at DEKA would get a fractional piece of Ginger, beginning with a hundred shares for secretaries and graduating up, for a total of 2 million shares.

"They've had an investment adviser and didn't know it," said Dean, "and it's about to pay off. I can't give them shares of DEKA. DEKA is my arms and legs and body. So shares in DEKA would never be any good, because they could never sell them to anybody."

Michael Schmertzler had objected to the dilution of CSFB's stock caused by giving shares to employees. "I said to him, 'Why do you think we *have* something to sell? Who do you think is going to make it and make the *next* thing?' So he backed off."

At Ginger the largest portions of shares, in descending order, would go to CEO and president Tim Adams, director of manufacturing Don Manvel, and director of marketing Mike Ferry. Doug Field, director of engineering, wasn't included in their category and I wondered why.

"Doug is clearly in a different zip code there," said Dean. "Ron Reich and Phil LeMay will probably get 5,000 or 10,000 shares to start because they're good engineers and they moved their families here. Don Manvel was running a half-billion-dollar company and was probably making $300,000 a year, with options and a car and a pension. We had to offer him more to get him. If Don gets 200,000 shares, where's Doug? 100,000? No, because he was already here, he didn't move his family, he doesn't have twenty-six years of experience. He got to move to a new project, he got a raise and more freedom and more funding. He should pay *me* to work on this project. But is 10,000 right? No. I can rationalize almost any number. That's why giving shares is fraught with land mines."

In the new plan, Dean retained 74 percent of the company, worth an estimated $370 million. The investors felt confident of a tenfold return within the first few years. If they were right, Dean's piece would soon be worth $3.7 billion.

While Dean was dangling the investors like marionettes, he didn't let Ginger's CEO and president see the strings, much less touch them. Tim Adams had overseen half a dozen large acquisitions while reconfiguring

Chrysler's business in Europe and was accustomed to making pricey deals. But Dean and Bob Tuttle didn't consult him about Ginger's financing or even include him in the loop. When Dean and Tuttle met with Sahlman and the angels, Tim wasn't invited. Tuttle did occasionally send him financial documents—as a courtesy, not for feedback. When Dean did mention the investors to Tim, it sounded as if he considered them lenders, not partners. Tim didn't even know when the funding would arrive or how much it would be. This uncertainty frustrated his planning and budgeting.

A year after coming to DEKA, Tim still felt himself riding a sharp learning curve whose centrifugal force often required a strong grip. He had understood within months that Dean wasn't going to let him run the company, at least not at this point. Dean would make the decisions. It was exasperating, but Tim wasn't the type to let his ego override his professionalism. Instead he became a student of Dean and the DEKA model. He soon realized that when Dean felt attacked or even pushed, he dug in. If Tim wanted his points heard, including his strongest certainties, he had to present them as possibilities and then give Dean time to examine them from all angles and tiptoe toward them. It sometimes made Tim look passive or intimidated, though he was neither, and he often chafed at the necessity to appear so, especially to the Ginger team. There was so much to do, but Tim, like the marketers, felt handcuffed.

People around DEKA often mocked the car companies as sluggish dinosaurs, and Dean considered himself a swashbuckling entrepreneur, but Tim was struck by how slow DEKA often moved compared to Chrysler. He had always given his managers the leeway to make quick decisions. Otherwise you wasted time running up and down the organizational chain, and in the car business that kind of micromanaging was disastrous. But that's how things were done at DEKA, where so many decisions, including many that Tim considered unimportant, had to be run past Dean, or you'd hear about it later.

On the other hand, Tim said, "There are things we can do here because of Dean that I'd never think of as even imaginable, because of the contacts he has." Like Dean's idea to lasso fifty dignitaries by giving them free Gingers. "In my past life I'd never *think* of pulling that off," said Tim, "but Dean will do it. He's a great salesman. When you put him in front of people and have him talk about his passion, it's very powerful and persuasive. Whereas if *I* said, 'We're going to change the world,' it wouldn't work." Tim had always defined a company as a group of coordinated teams in which no one was irreplaceable. But Ginger had an irreplaceable resource—Dean.

Yet Tim and Dean clashed about fundamental things that could make or break the business. Most of them related to Dean's unfamiliarity with large-scale manufacturing. Nevertheless, he had strong opinions, or perhaps instincts, about it. Part of Tim's job was to protect Dean from himself. To build a big business quickly, as Ginger's business plan called for, Tim needed to hire a core of experienced people. But experience was expensive, and "expensive" gave Dean hives. He had vigorously resisted giving Don Manvel the salary necessary to lure him as director of manufacturing operations.

When Tim told Dean they needed a good procurement guy, who might cost $200,000, Dean retorted that Tim was nuts, that he should hire someone smart for $35,000 and train him. Dean had built DEKA using that philosophy. Tim understood that. But DEKA wasn't a large manufacturing business requiring millions of dollars in hard costs. Once the new company was up and running, there would be lots of room for smart young people. But when you were hurtling toward a launch date, you didn't want someone figuring things out as he went along, because he was going to make mistakes that could cost hundreds of thousands of dollars. You needed people who understood highly repeatable operations, high-volume production, and high-volume testing, not to mention comprehensive sales, service, and distribution. So when Dean asked, "Doesn't a procurement guy just write up orders?" Tim's answer was no, he manages supplier relationships so that when problems come up, they're fixed before they kill you. But Dean resisted, always looking for ways to squeeze nickels. It slowed things down and delayed the plan, but of course Dean didn't want to hear that.

Tim had been telling Dean for months that they needed to make a decision about space for a manufacturing plant and offices. Tim wanted Ginger to be separate from DEKA. So did Doug Field, Don Manvel, and Mike Ferry. Tim sensed that the culture developing at Ginger wouldn't mix well with the DEKA culture. DEKA ran fast and loose, an R&D group with little structure, whereas Tim and Doug and Don preferred to operate with well-marked boundaries such as deadlines, budgets, sign-offs, and strict accountability—necessities for a large manufacturing company.

The Ginger team also wanted to be independent, which was impossible under Dean's eagle eye. And they thought it was crucial to keep the entire team together. Engineering needed to stay with manufacturing so that assembly problems caused by the design could be fixed immediately. Mike Ferry wanted to keep the marketing group close to Doug's team so the engineers didn't forget consumers. The obvious solution was a manufacturing facility with office space for everybody.

Dean's indecision about a permanent location had started to affect morale. Team members were asking why they hadn't yet cut loose from DEKA. Tim and Don had been scouting sites within thirty miles of Manchester and presenting the options to Dean. But Dean always found a reason to veto the sites—too expensive, too far away, too one-thing-or-another. He occasionally implied that he might want to keep Ginger in the Millyard, snug next to DEKA. Tim and Don shook their heads at Dean's naïveté about high-volume manufacturing. They needed at least ten loading docks and easy access for fleets of tractor-trailers, not to mention parking for two hundred fifty cars. None of that was possible within the Millyard's narrow lanes and dense layout. Tim and Don also wanted a flexible, "breathing" plant—a big one-story box that could be adapted for different manufacturing lines and volume flows. That too was impossible in the Millyard's old-fashioned multistory architecture.

Tim had hoped to move Ginger to its permanent factory site by June 2000 so that the engineering builds after D1 could be assembly-tested there. That would have required a decision in February. Now it was May and Dean was still stalling. Tim kept telling him that the schedule was getting dangerously compressed, that in terms of risk management they were right on the edge. He warned that waiting too long could force them to accept whatever was available at the last moment.

But Dean didn't budge. Ginger's managers had realized that he couldn't bring himself to let Ginger leave home. He might permit the factory to be off-premises, but not the engineers. They were his, no matter what the company's name was.

Dean's reluctance to pull the trigger chagrined Tim. One reason he had loved the car industry was that the bets were so huge, the competition so fierce and fast-paced. He looked forward to that phase of Ginger and resigned himself once again to patience. Meanwhile, he often felt out of his element. Dean seemed to focus simultaneously on minutiae and the grand vision, whereas Tim operated best in the territory between the two. They could make a great team, if Dean would ever trust him with the ball.

When Dean was in California visiting Jobs, he had met Mike Ferry, his director of marketing, at Lexicon's headquarters in Sausalito. Lexicon designed names for new companies and products. This process, called "branding," was serious, expensive business. Lexicon was well known for creating names that sounded cool or techno but had no meaning in English. Most people call such sounds gibberish, but Lexicon presented them as "empty vessels" that their clients could fill however they chose to.

The consumer landscape was littered with Lexicon's neologistic rubbish: Pentium and Celeron, Zima and Zeprexa, Dryel and Febreze. Their linguists preferred some meaningless sounds over others: Viant, Ventro, Verado, Vistide; Nuon, Xeon, Luxeon. Lexicon also specialized in names that evoked something-or-other, but conveyed no discernible connection to the product: Wingspan (a bank), Recode (clothing), Blackberry (a wireless e-mail device), Blue Nile (an Internet diamond company), blueyonder (an Internet service). Occasionally the company hit something dead-on, but for every Apple Powerbook or Subaru Outback there were a dozen Amphires and Cellegys and Procurons.

The Lexicon team, including founder David Placek, gave Dean and Mike Ferry twenty names at a time, over ten rounds. The exercise alternately amused and annoyed Dean. He pissed into Lexicon's empty vessels and blew away the vague evocations. Blue Wagon, in particular, inspired a stream of sarcasm. In his view, expressed repeatedly, the message of such names was, "We couldn't think of a name ourselves so we paid some clowns $70,000 to do it."

The Lexicon people, artists and scholars of branding, were a bit affronted, and Mike Ferry was embarrassed. Dean didn't care. He wanted a name that evoked his particular machine and that would let him tell a story. The name also had to sound sidewalk-friendly. That killed Zume, among others. He sort of liked Relay and Standing Wave, terms from engineering and physics. He wanted to think more about Transer and Circa. So far he preferred his own newest suggestion: Rev, as in Revolving, Revision, Revolution. He could tell stories about all those things. Ginger was his baby. His name and reputation would be attached to it. He would decide what to call it. For the time being, the company would keep operating under the placeholder name of Acros.

Up in the Penthouse, the engineers were busy at the D1 "raft." This was a large piece of plywood on which Ginger's parts were laid out side by side and connected with wires—boards, motors, batteries, transmissions, gyros. The horizontal configuration made it easy for the engineers to test parts and code.

"Shoot some code into this chip for me," said Phil LeMay to a software engineer, who began typing at a keyboard. The code made the motors spin. "Something smells funny," said Phil, sniffing. Burned plastic. Phil applied a diagnostic tool—his nose—and snuffled up and down the board like a bloodhound.

A few days later Doug was listening to the motors and transmissions with half-closed eyes. In C1 the transmissions had sounded like banshees. These were better. "There are multiple tones that should move up and down harmonically with the speed, like music," said Doug. "It's the audio portion of Ginger." He had picked out the tones he liked from a compact disc supplied by the manufacturer. The disc contained sound samples made by various gear teeth, which had to be "tuned" together. "There are excitation frequencies and response frequencies," said Doug, "and as these frequencies go up through the machine they create *more* frequencies." Doug was the conductor of Ginger's conductivity.

It was mid-May now, two weeks until the D1 deadline, and some of the engineers had started wearing shorts and sandals to work. Pairs and triplets were staying until midnight, 2:00 A.M., 5:00 A.M., to finish their pieces of the machine. "Back when everything was square," said one exhausted worker, "engineers used to go home early." A part was not considered finished until every engineer whose parts it affected had signed off on the drawing.

One afternoon J. D. Heinzmann sat at the raft, wearing a big grin because his redesigned boards were working so well, with less noise and interference. Not that there weren't still problems. He was running torque/speed curves on the oscilloscope, and the waves looked as jagged as the Tetons. "That's *really* ugly," he said. They were supposed to be rounded curves.

Jim Dattolo sat across the lab table from J. D., providing code. "Reality ends here," said Jim, pointing to the midline between them. A software engineer, Jim was by definition immersed in unreality. Regular engineers were suspicious of software jockeys because their work didn't obey the laws of physics. Code didn't melt at a certain temperature or emit a measurable wavelength. Software engineers created fictions in mysterious languages, and then piled on more fictions until they had constructed an entire world, all of it invisible and make-believe. And yet this make-believe commanded the motors and gyros and circuits and boards. It was enough to give regular engineers the heebie-jeebies. The two engineering camps, hardware and software, often blamed each other for glitches.

"His code is bloated," said J. D., still peering at his jagged waves.

Jim snorted. He had been lured to Ginger by Phil LeMay from the company where both of them had worked in Michigan. In his twenties, with a long ponytail and a thin goatee, he looked more like a musician than an engineer, and in fact played drums and wind instruments. When he focused, he was a dazzling coder, fast and intuitive. He viewed Ginger as a software problem in a cool box, and there was no doubt which component was more important. "The software will determine whether or not it's a great machine," he said.

It took a lot of typing to code-in functionality. Just to drive the motors, Jim had written more than a hundred pages of code, all of which went into a little black chip and got executed ten thousand times every second. He could talk about his work in striking imagery. "Code has to evolve or die." "Sometimes you have to do code surgery, but there's a difference between removing an organ and messing with someone's DNA." "For Microsoft, a bug is Control-Alt-Delete. For us, it's your face scraping across the pavement."

The project's other code designer was John Morrell, the lead controls engineer in charge of safety and vehicle dynamics. His responsibility was to ensure that Ginger never went haywire and stayed balanced under all circumstances. Jim Dattolo called this "the high-level stuff. John is the intelligence, I provide the raw power. If he asks for 50 volts, I give it to him."

Jim informed John about any change he made, no matter how

small, so that John could make sure their code still "shook hands." Every change was a patch in the code, and every patch was a weak point. They patched constantly.

"The curse of software," said John, "is that unlike the hard tooling, you can change it. And so people do, all the way to the end. But software is really brittle—it breaks easily, like glass."

"But brittle failures are easy to find," said Jim.

John looked amused. "That's the difference between a programmer and a controller," he said. "I'd rather have a tick-tick-tick in the machine than a brittle failure that stops the wheels and results in a faceplant."

"No," said Jim, "that tick-tick-tick drives me crazy. Those things *suck*. They're the ghosts in the system."

"Well, those are the problems that would drive our service department crazy," said John, "but the brittle failures would drive our legal department crazy."

After patching together the code for Ginger, they intended to remove the brittle areas by rewriting the entire thing. It had to be right. If a glitch in the code forced the recall of a million Gingers, the company wouldn't survive. But there was one potential failure they couldn't design out—a rider's lack of common sense. "Somebody probably *will* go over a cliff on one," said Jim. His job was to make sure it wasn't because of a software error.

But first things first. At one of the daily morning meetings in mid-May, Doug announced that they were finished with the raft. It was time to hand the baton to John Morrell so he could get D1 up and balancing. John nodded sharply, once. He looked keyed up, a Ferrari waiting for the light to change. He asked people not to give him feedback about the machine for a while. "I have too much onion to peel and I'm already crying."

John had a small room to himself in the Penthouse. He spent hours in there with a skeletal Ginger that sat on a milk crate, the machine's wheels spinning as he stared at lines of code racing across his computer screen. Sometimes he typed furiously and then studied the effect on the wheels and the screen. Periodically he would burst out of the room looking agitated or elated by his latest fictions. "You need long chunks of quiet time," he said, "because you need to build an absolute argument."

He had come to DEKA to work on Fred in 1996 after finishing his doctorate at MIT, where he also captained the bicycle racing team. For two years the work at Fred exhilarated him. "Dean is good at expanding choices," he said. "At that point he's an incredibly exciting person to work with." But by the time Ginger opened up, Fred had bogged down and John wanted out. Doug poached him for his team. Doug considered him brilliant. Watching him at the keyboard, said Doug, was like watching a great concert pianist.

John's personality combined witty high spirits with philosophical introspection; he had to balance those as well. He probably had been thrown from Ginger more times than anyone, with the possible exception of Doug, because he was his own guinea pig for testing the controls code that regulated the machine's behavior, or failed to. He would fall off, punch in more safety code, then get back on and repeat the cycle. "There's so much uncertainty across so many dimensions," he said. "It makes for a lot of pain." He often rode with only one foot on the machine so he could jump off fast.

Other team members found Ginger's temper tantrums amusing. Not John. He knew just how dangerous they could be, and the responsibility to fix them was on his shoulders. The machine scared him. So did the idea that someone might get hurt because of a flaw in his code. "It's not clear that a self-balancing machine is really a good idea," he said in one of his gloomy moments. "I still sometimes think that we're fighting the gods."

But over the next few days he won the battle, if not the war. On Friday afternoon, May 19, the team called Doug, who was en route back from a business trip, to tell him that "Ginger Fields" was up and balancing for the first time. Doug could hear their cheers in the background. The machine was jittery and tenuous, like a baby learning to stand, and it couldn't yet move. Four nights later John and some others called Doug at home at 11:00 to report that D1 was "walking, running, and dancing."

"What's left is to dress her up for her coming-out party," said Doug.

That's what they did in the last few days before the deadline. They carved bits off the battery packs so the packs would fit into the new chassis. They shaved steel off the bottom of the control shaft. They checked and rechecked DEKA's shop for freshly machined parts. At one point when something didn't fit, Phil LeMay released his stress by walking over to a fan hanging in a doorway and humming into it, altering his tone to get different frequencies.

The battery charger finally arrived from the manufacturer, Saft, on May 29, weeks late. It didn't work. The charger was Bill Arling's part. Saft had been driving him insane with broken promises and bad information. He retreated to the hanging hammock-chair near his desk, put on his headphones, and closed his eyes.

I asked Doug if they would still meet the deadline. "Of course," he said. "Horton the Elephant is my hero: 'I meant what I said and I said what I meant. An elephant's word is 100 percent.'"

People were crimping, drilling, filing, soldering. They made a seal for the dashboard module by cutting up a surgical glove. They bought emergency parts at Radio Shack. Kurt Heinzmann wandered in from the iBOT project. Kurt had been the first engineer to stand on the little balancing platform that turned into Ginger. He peered into the new chassis: "Wow! So few wires!" He kept wowing as Doug showed him the new wheels, handlebar, and control shaft. He ran his hands along all these parts with a wistful expression, like a parent who hadn't touched his child for many years. A row of older Gingers lined one wall. "Lots of evolution there," he said fondly.

The machine began coming together. Batteries and boards into the chassis, control shaft into the base, dashboard into the handlebar, handlebar onto the control shaft. The fenders wouldn't arrive until tomorrow, but the machine was starting to look beautiful. It was strikingly different from earlier versions. Everyone gathered around and stared.

But would it work? Someone flipped the On switch.

Nothing.

Several hours later, after a lot of hovering and disassembling and diagnosing and tinkering, the LED lights finally blinked on, anticlimactically. But many things needed to be checked and rechecked. They kept taking the machine apart, making adjustments, finding new problems. The next morning, May 30, no current was reaching the board beneath the dash. They disassembled the dash and the control-shaft base for the umpteenth time and discovered a bent pin on the connector. I remarked how fragile the machine seemed, if one pin could shut it down. "If a pin *couldn't* shut it down," said Doug, "why would we need that pin?"

They had saved the wet test for last. Doug stood on a stool and poured water over D1. John Morrell flipped the switch to turn it back on.

Nothing.

"We've got some work to do," said Doug, grinning. They found water in the chassis, the control shaft, and the dash. They started looking for the leak, or leaks. Two engineers carved a new rubber O-ring so big that it bulged from the midsection of the control shaft. They doused the shaft in the sink and then one of them unscrewed the connection and peered into the shaft with a flashlight. Time passed. Bill Arling, the second engineer, finally exploded: "*WELL?*" More time passed. "I can't *stand* it," said Bill, stalking off. "It looks dry," said the first one, laughing. Problem solved.

Doug had been pacing around the Penthouse in a way that managed to seem simultaneously patient and impatient, head high, brow furrowed. Now he noticed the bulging O-ring. "That seal is highly visible to the user," he said. "It has to be trimmed or otherwise solved. It violates one of our design principles, which is that the basic form of D1 should be correct, and that isn't." The engineers removed the shaft again, to do some surgery with a razor.

When the fenders arrived at 5:00, Scott Waters held one up to a wheel and groaned with pleasure. "That is the *balls!*" said John Morrell. They tested the jury-rigged charger. It didn't work. Doug asked Bill Arling what his plan was. "To stay here and figure it out," said Bill, looking glum.

At 6:00 Doug gathered the team to assign roles for tomorrow's demonstration of D1 to Dean and Tim. A few small things needed doing, plus the charger. "So we all agree," he said, "that tomorrow morning we'll have a working machine and a working charger unless I hear differently tonight. Right, Bill?"

"I need to get to *work*," said Bill. His eyes were intense enough to solder metal.

"It could be worse," said Phil LeMay.

"It could barely be better," said Doug. "Considering all the risks stacked on top of each other, the twenty or thirty key technical changes we made to this machine, it's amazing."

That same evening, Dean was telling Tim Adams his latest design idea: a handlebar that would be more pedestrian-friendly. That way, a bumped pedestrian wouldn't take a full hit, because the soft bar would absorb some of the energy.

Tim liked the idea but didn't want the team to get Deaned at the very moment they were trying to finish D1 for tomorrow's deadline. The idea also would require a redesign of everything above the chassis, and it was too late in the launch plan for that. He and Doug had been pushing Dean to let an engineering team split off to work on the next generations of Ginger, but so far Dean had balked. Now Tim said that Doug's guys were too strained at the moment, but the idea was worth considering for future generations.

Dean walked over to Ginger. Most people had gone home, their jobs done for tomorrow. Industrial designers Scott Waters and Tao Chang were painting the fenders and doing touch-ups. Doug proudly put a fender on for Dean. "But it needs to come out over the tire," said Dean, studying it. "Otherwise the tire will leave a mark when it hits a wall and leave dirt when it hits a pedestrian. I wonder why Scott didn't cover the tires?"

He also wanted the handlebar changed. The current design formed a loose W. Dean didn't like the pointy bottom ends. He wanted them rounded, to be less hazardous for riders and pedestrians. Doug and Scott immediately understood and agreed.

When I left Ginger at 12:30, Doug and the industrial designers were still primping D1 for the debut tomorrow.

At the morning meeting, Doug said he had seen Ginger in her wedding dress late last night, and she was beautiful. He knew that everybody already had changes in mind, but today was the honeymoon and criticism was off limits. The exhibition was set for 4:00 in an empty room downstairs.

Most of the team spent the day relaxing. Bill Arling, the charger solved, wrapped new tape around his bike's handlebar. Tao Chang dabbed the tires with a black marker. Even J. D. Heinzmann, he of the sixteen-hour days, was sitting around.

Dean marched into Ginger at 3:55, straight to Doug and Scott Waters. Everyone else was already downstairs. "This probably isn't the best time to mention this," said Dean, and then he launched his riff about the new idea for the handlebar. Doug looked dismayed. When

Dean started sketching his idea, dismay turned to irritation. Getting Deaned was not what Doug wanted today, especially not right now.

"We have to be downstairs in two minutes," he said. The hint bounced off Dean, who kept riffing and sketching without a pause. Doug's face grew more and more stony. "I have to go downstairs," he said gruffly. "People are waiting." He turned away and left. Dean paused, surprised, then revved back up and resumed Deaning Scott Waters, who wore a frozen grin.

Dean showed up downstairs ten minutes later. Balloons decorated the room. In the middle of it, a black sheet covered D1. Doug ceremoniously lifted it and everyone applauded. "I must say," said Doug, "it's one of the most beautiful sights I've ever seen."

He went through the checklist of goals for D1, putting the machine through tasks and functions and modes. He clicked a stopwatch to show how quickly Ginger reached operating mode after being turned on: 1.6 seconds. To demonstrate smoothness, he put a cup of water on the platform and rolled the machine: no spillage. An engineer took it up a 3-inch curb and off of a 6-inch jump. Finally Doug showered the machine with a gallon of water, then asked John Morrell to restart it. Teeth clenched, John flipped the switch. After an agonizing fraction of silence, we heard a sweet little beep. The team cheered.

Dean stepped onto the new new Ginger. He looked pleased. "I have no doubt at all that this will revolutionize the world," he said, spinning in slow circles. He told them that Ginger would change city planning and people's lives. "The impact of this in the twenty-first century will be just like what Henry Ford did at the beginning of the twentieth century."

Still spinning, he told them that Bob Tuttle was in New York at that very moment signing for nearly $80 million from Credit Suisse First Boston and Kleiner Perkins. It was KP's largest deal ever. He explained how it had happened: "I did it the way I usually do: I figure out what I think is fair and then make sure everyone compromises and does it my way. We will control the company. We will control the board. We will control the vision."

He went into prophetic preacher mode. Ginger would be the main form of transportation in some cities within five years, as common as cars in ten. "Think of the major inventions of the last century," he said, slowly spinning, "the car, the radio, the TV, the computer. I'd say this is in the same category." He let that sink in. "Congratulations to everybody," he said. "This machine is awesome."

The team basked in the glow of Dean's vision. What group wouldn't feel swollen by such praise from their leader? John Morrell once said that when Dean expressed his pride and satisfaction, "You could *live* off that." And then, always self-conscious, John had smiled at himself and added, "We're basically a cult. All the characteristics are there—a charismatic leader and devoted followers who sacrifice their time and money for him."

At this moment Dean's low salaries and sharp demands, his cracks about stupid charts and wasted money, his habit of Deaning them at bad moments—all of it was washed away. What mattered was that D1 was finished, Dean was riding it with pleasure, and the future he envisioned for them was spectacular.

The next day, the engineering team took a celebratory paddling trip down the tranquil Merrimack River. It was a perfect June morning. A couple of engineers brought their own kayaks, but most of us partnered in rented canoes. John Morrell made the trip in his strange boat, a slalom open racing canoe. That awkward phrase matched the craft's behavior when anyone inexperienced tried to paddle it. J. D. Heinzmann, a skilled canoeist, took John's boat out for a spin while the rest of us ate lunch on the riverbank. When J. D. could make the boat move at all, it zigzagged like a crazy pattern on an oscilloscope. He had to paddle hard to return to shore, where he panted that the boat was almost impossible to turn or control.

Then John Morrell got back in and made it twirl and sprint and caper. "That's magic!" shouted J. D. Naturally the team's dynamics expert knew how to control the fleet's most unstable boat.

The canoe carrying the two youngest engineers was dubbed "the frat boat" because of its steady discharge of boyish energy via splashing, water balloons, and general prankishness. We came upon a rope swing on the bank and took turns swooping out over the river and dropping off into the cool water. When Doug Field tried it, he slipped down the rope and didn't lift his legs, so his long body dragged through the water like a sack of potatoes. It was the only time I ever saw him look graceless.

Some of the guys tried to dive after letting go of the rope, but no one succeeded. They began working out the physics behind their failures. The same spirit probably piqued young Galileo as he mused about the math traced by the swinging lamps in Pisa's cathedral. Galileo believed that scientific observation was the most trustworthy path to knowledge. Those pendulous lamps eventually led him to capsize Aristotelian physics and discover new laws of motion and dynamics.

Here on a New Hampshire riverbank, Galileo's descendants broke down today's equation: the length of the rope, the angle of the bank relative to the plane of the river, the arc and period of the pendulum, the weight of the swinger.

J. D. suffered for his theories with several spectacular failures, smacking the water so hard that his back and belly resembled poached salmon. But he persisted toward a solution. Hard experience taught him that he needed to lengthen the arc and its period, to give him more control over his momentum and center of gravity. Using the tools and technology at hand, he asked two engineers to lift him so he could stand on the big knot normally used as a handhold. They pulled him a bit farther up the bank and released him. He swung out over the river, farther and higher than before, and when he let go at the end of the arc, he reversed his poles to complete a graceful dive. He came up whooping, with that joyous grin that made him look like an overgrown kid.

Phil LeMay, smoking a cigar, applauded. Phil was sweating but didn't want to swim, so he was standing calf-deep in the brisk river. "To cool off, you put your radiators in cold water," he explained. "If you get cold, you wiggle your toes for thermal conductivity."

Engineers. Where I saw scenes with literary and painterly overtones, they saw physics puzzles and thermodynamics. We both loved the physical world, but absorbed and processed it differently. "I'm perfectly willing to watch a sunset," Dean once said, "but I also want to know why it's orange and why it looks bigger when it touches the horizon. People think science is dry," he added, "but it's as emotionally charged as looking at a beautiful woman. The most shallow bunch of pinheads I know are arts people. I ask them, 'So you're cultured and well read? Tell me one differential equation you know, one piece of elegant physics.'"

Some Fred engineers had once taken a canoe trip on the Merrimack. Dean had surprised them in his helicopter and followed them for more than an hour, swooping down to churn up a prop wash that wobbled their canoes. Today the Ginger team half-expected him to skim over the treetops, the deus ex machina that sometimes resolved problems, sometimes delivered them. Literally or otherwise, he was always a hovering presence. But except for sporadic bombardments from the frat boat, the river stayed calm and quiet.

As we paddled, one team member mentioned to another that at D1's debut yesterday, Dean had said that he intended to be around more after the financing was settled. They traded uh-oh looks, the fraternity of the Deaned. "If he is, it will affect our deadlines," said one of them. "We could chart it with a scatter graph, plotting the number of Dean's visits against the project goals."

Engineers. They laughed and glided on, toward the rapids in Manchester.

The Slip

June 2000

The night after D1's debut, John Doerr called Dean from India. Kleiner Perkins wanted to be the lead investor on the Stirling engine project. Earlier in the day, Dean had spent two hours on the phone with Michael Schmertzler from Credit Suisse First Boston. Schmertzler insisted on being equal partners with Doerr on the Stirling, the same consideration that Credit Suisse had given to Kleiner Perkins on Ginger. But Doerr didn't want Schmertzler to have any of the Stirling.

Dean wanted to keep Schmertzler happy, but Doerr was adamant. Dean worried that if Schmertzler didn't get equal treatment on the Stirling, he might demand that Doerr be expelled from the Ginger deal. Dean hoped the banker didn't play that card, because he didn't want to play his answering card: "I guess we'll have to let Kleiner Perkins and Jobs have it all."

He looked exasperated. We were in his library. Dean turned to the laptop that stayed open on his desk and launched a CAD program. "The more I look at the handlebar," he muttered, "the worse it gets." He began drawing on the screen. "It shouldn't even *be* a handlebar. That's for a bike or a motorcycle, not something on a sidewalk. It should be small." He pulled a tape measure from his pocket and measured his palm. "You only need 4 inches." He turned back to the computer.

As he clicked and drew, he worried aloud about one of his engineers whose four-month-old baby was ill. The infant's skull had closed too quickly, putting pressure on the brain. Dean had called a contact at Harvard Medical School and gotten the name of the best surgeon for the required procedure. The baby had been operated on today in Boston. Dean was wondering what had happened and how his engineer was doing. "If he's freaked out, I'll have to go there tomorrow," he said, drawing different shapes for the handlebar, worrying about his investors and his regulators and his people, multitasking.

A couple of weeks later, in mid-June, Dean caught me up on things over Chinese takeout in the mall's food court. Both Schmertzler and Doerr had told him that Ginger's entire management—CEO Tim Adams, director of engineering Doug Field, director of marketing Mike Ferry, and director of manufacturing and operations Don Manvel—weren't up to their jobs. Their opinions shook him. He liked to keep his investors happy and confident in him. He told them they were wrong about the team, but to me he worried, especially about Tim and Mike. His investors constantly seeded him with doubts about the team's leaders. Some of them wouldn't survive it.

The previous week Dean had fulfilled a promise to himself by taking the entire iBOT group to a trade show in Paris. "I took the machine down the Champs, down the stairs into the Metro, back up the stairs to the street, and up the stairs to the restaurant in the Eiffel Tower, where I hosted a dinner for ninety people." He looked pleased, remembering it.

"You should have been here yesterday," he continued. On Bill Sahlman's recommendation, Joy Covey had come to visit. Covey had studied with Sahlman at Harvard and eventually became chief financial officer for an obscure start-up called Amazon.com. She helped Amazon's founder and CEO Jeff Bezos raise the money to make the company the most powerful retailer on the Internet. Still in her thirties, she was rich with Amazon options but recently had left the company and was looking for something else to be passionate about.

According to Dean, she had found it yesterday at DEKA. Since Ginger was spoken for, she wanted to come in on the Stirling project and arrange the financing. Once again, Dean was amazed by the speedball attitude of dot-com people. But that wasn't the way he worked. If she really wanted to jump into something, he told her, she should read this literature about FIRST.

Michael Schmertzler from CSFB had called this morning, ripping mad because John Doerr had told him that Dean was going to let KP be lead investor on the Stirling. Doerr seemed to enjoy tormenting the bankers. Dean hadn't promised anyone anything about the Stirling. Doerr also was equivocating about signing the papers for the Ginger deal. Dean shook his head.

"They're like kids trying to push each other around," he said. "If they knew how desperately we need money right now—we're running out. But all they see is how much Ginger has progressed, and everything that's being written about water and electricity and transportation." He paused. "I think the dot-coms are to the point where they realize that somebody had better start making *products*."

He had a new favorite name for his: Flywheel. The marketers liked it because of its evocative combination of words, but to Dean it meant much more. "It's an engineering term and it's one of the concepts that started it *all*—a mechanism that makes something go. It's perfect. I couldn't believe I hadn't thought of it. What's the first thing you see when you walk into my house? Those huge flywheels."

Dean had literally built his house around two massive antique flywheels. Weighing 8,000 pounds each, they dominated the entry, rising 20 feet from the basement through the first floor. An enormous steam engine still sat between them. Dean liked to envision the flywheels spinning in the middle of his house, making power. He had been renovating the mechanism for years, making replacement parts for it at DEKA or in his home shop. I once saw two gigantic steel nuts sitting on his desk. A grapefruit could have passed through their holes. Dean was going to put them on the flywheels that night. I wondered where he had found a wrench big enough for the job. "I made one," he said.

There was one obstacle to calling the machine Flywheel: Someone already owned the domain name "flywheel.com." Dean hoped to buy it away.

It was time to finalize Ginger's board of directors. Sitting over the remains of Chinese glop in Styrofoam dishes, Dean ticked off the names: John Doerr, Michael Schmertzler, Bob Tuttle, Vern Loucks, Paul Allaire, maybe Joy Covey. Steve Jobs, unofficially. He started laughing. "And *me!*" It still seemed preposterous to him to be in such company.

After dinner he wanted ice cream. He rarely drank coffee and only occasionally drank beer or wine, but he never passed up a chance for ice cream. The teenage girl behind the counter asked which kind of cone he wanted. They offered three types, she said, pointing.

"The conical one," said Dean. She looked blank, so Dean tried again: "The one that's a *real* cone."

Her eyes flickered with mild panic. Guessing, she pointed to one of the cones, which had a cylindrical bottom and a larger cylindrical top.

"No," said Dean, baffled by her confusion. "The *conical* one." He couldn't be any more clear. Things like this confirmed his fears about the technological illiteracy of American youth.

The girl looked flustered and embarrassed. I knew how she felt. Though I understood every word the engineers spoke, they sometimes combined words in phrases and concepts that left me feeling lost in Babel—"There's a 50-hertz anti-alias pull on the motherboard," or "Convolution in the time domain is multiplication in the frequency domain."

"The sugar cone," I told the waitress, translating geometry into the vernacular.

"Oh!" she said, her face relaxing. "The *sugar* cone!"

And that's how Dean finally got his ice cream.

Like Dean, industrial designers Scott Waters and Tao Chang were dissatisfied with D1's handlebar. They were discussing how to deal with the bar's bumpers, essentially the plugs on each end. At the moment, the bumper was a round black rubber plug with a screw hole in the center. Tao had sketched three dozen alternative possibilities on yellow paper and now he and Scott pored over them, shaving an angle or extending a curve. They often paused to study the images.

"But there's something nice about the simple round shape that's there now," said Scott.

Tao smiled. "But it looks like a door knob."

"I know that's what someone said," said Scott, "but I don't think so." He sighed and circled several sketches. "If we're going to move away from a circle, I think we should *push* away from it, not tiptoe."

It seemed like a lot of bother for such a minor detail. "It's just a bumper," said Tao, "but it's hard to get it right. People don't believe how long it takes. They won't even notice if everything works well together, but if a part *doesn't* fit, they notice right away."

To respond to Dean's concerns about D1's fenders, Doug, Scott, and a couple of others had taken the machine to Dean's house for testing. They put an inch of water on the driveway and rode through it to measure splashes. Then they taped bits of cardboard onto the fenders to determine how much of the wheel they needed to cover. By adding just an inch to the fenders, they could roll through puddles at full speed, stay dry, and kick up an inconsequential rooster tail.

That relieved Scott. The tires and wheels were Ginger's main visual components, its great gams, and he wanted to flaunt them, not cover them with dowdy full-length fenders as Dean had suggested.

While they were splashing through puddles, a helicopter approached. When they realized it wasn't Dean, Doug hastily covered the machine. The chopper veered off. They started laughing about Doug's paranoia. Someone joked about seeing a big H on the chopper, for Honda, and hearing the clicking of high-speed cameras. But if the helicopter had returned, they would have rushed to hide Ginger again. Dean had taught them well.

Within a week of finishing the one-of-a-kind prototype, D1A, the engineering team had compiled a long list of small changes for the machine's next quick revision, D1B. The team expected to build forty or fifty D1Bs, incorporating things they had learned. The longer-term revision would be called D2 and would require a full redesign of the charger, the kickstand, and especially the dashboard, the component that concerned them most. They also needed to revamp the embedded electronics and software. They had to refine the clamping on the control shaft. They weren't yet confident about sealing or electromagnetic interference. They needed to reduce cost and weight. And, of course, none of these components had been rigorously tested yet. All of this would eat up months.

Another foggy, scary category loomed: "regulatory requirements." It had begun to dawn on Dean and Tim Adams, and hence on the engineers, that they didn't know how different states and foreign countries regulated motorized machines. For Dean and Tim, the main issue was getting Ginger on the sidewalk, but for the engineers the issue was design. Some governments might require brakes or horns or reflectors or turn signals. The possibilities were as endless as bureaucratic fussiness. No one had bushwhacked into those regulatory thickets to see what might entangle Ginger, and the engineers needed to know before freezing the design and tooling up for manufacture. Otherwise the changes could be financially disastrous.

In mid-June the engineering team leaders gathered in the War Room to discuss all this. Then Doug moved to the final issue on the agenda. He wanted to revisit the overall schedule. He reminded everyone to be circumspect about repeating anything to the rest of the team, because they didn't want to alarm people. "Part of our function is to provide damping of those oscillations." And since the words "months" and "years" carried "emotional baggage," Doug wanted to reestimate the schedule using numbers, working back from 0, which stood for First Salable Machine, then back through Revalidate Design Changes, Change & Reprocure Parts, Finalize Design, Vehicle Test Program, V1 Build, Long-Range Fixes, Component Testing, D2 Design Changes, D2 Testing, and D2 Build.

He was dancing around the topic, introducing it to death, which wasn't like him. The schedule was certain to be discussed at the upcoming July board meeting, he continued. "And if we don't come up with the timing, someone else will, and that's much less pleasant," he said. "My philosophy is that when you make a Gant chart, you have to guess, and then you make that guess come true."

So they started guessing. Doug drew a chart on the whiteboard and moved backwards, right to left, as they estimated the project's remaining phases in blocks of months—or rather in numbers, which everyone immediately translated into months. The current plan called for launching Ginger less than eleven months from today. But as the engineers in the War Room predicted how long they really expected things to take, Doug's chart quickly backed up beyond block 11. When they finished, they had reached 22.

Everyone stared, stunned. Ron Reich began talking to himself in a low moan. Someone broke the tension by asking if he was having a nightmare. But it was no joke. They studied the chart again, more intently, and eventually shaved three months.

"So, nineteen months," said Doug, "which puts us at September 2001. That's about what I expected."

"Wait," said Phil LeMay. "Nineteen months is January 2002." To be precise, the end of January.

"That doesn't sound good," said Ron.

Doug rechecked the figures. Phil was right.

Time added to a schedule is called a slip. This was a big one.

Near the end of June, Dean went to a gathering in Aspen hosted by Kleiner Perkins for the CEOs of the companies KP had invested in. There, an enthusiastic man with lively, intelligent eyes greeted Dean like an old friend. Dean didn't recognize him and felt embarrassed when the man reintroduced himself: Jeff Bezos, founder and CEO of Amazon. Bezos had met Dean at one of the TED conferences. Their mutual connection was John Doerr, who had funded Amazon.

Bezos was a good listener and knew Doerr well, so Dean told him about his troubles. Doerr had agreed to equal terms with Credit Suisse First Boston, but was still trying to tweak the deal. Unlike CSFB, KP hadn't signed the papers. Dean was exasperated and so was CSFB's Michael Schmertzler. Bezos sided with Dean and offered to talk to Doerr for him. In gratitude, Dean invited him to see something special in his hotel room—Ginger.

The machine amazed Bezos. He rode it all around Dean's room and said he *had* to get involved. Dean liked him. He was sharp and enterprising, with no airs, a refreshing quality in a billionaire.

Dean told me all this in his helicopter on the way to pick up Bezos at the Manchester airport. Bezos had called last night and invited himself to DEKA. Who could refuse? People considered him a marketing and retail genius. Long before Aspen, Dean had marveled at Amazon's easy Web interface and customer service. He once grumbled that every time he logged on to Amazon, he spent more money than he intended because the experience was so pleasant. He had already told the marketing team that Ginger's Web site and customer service should be modeled on Amazon's. Now he had a chance to pick the guy's brain.

So Dean had cancelled tonight's dinner appointment with Michael Schmertzler, citing Bezos's visit. He wondered whether Schmertzler suspected Bezos of being a fifth columnist for the conspiratorial Doerr. Dean also worried that Doerr might suspect him of courting Bezos to replace KP.

Bezos's jet, a chartered Gulfstream V that comfortably seated more than a dozen passengers, today carried just one. "Dean!" said Bezos as he bounced down the staircase. Bezos stared at the helicopter and said he had never been in one because they terrified him, but he supposed that flying with the guy who designed it and owned the company might be OK. Once aloft, he loved the ride and expressed his pleasure with his trademark honking laugh, which would make an effective car alarm. He told Dean that when his siblings were teenagers, they refused to go to the movies with him because the laugh embarrassed them. Dean put on the *Star Wars* CD, which made Bezos laugh harder. Dean swung over his hilltop mansion before heading to DEKA. As the neat brick rows of the Millyard came into view, Dean mentioned that people had called him crazy for buying almost a million square feet of run-down factories.

"When they call you crazy," said Bezos, "you know you have a good idea."

"Right," said Dean. "If they say, 'Hmmm, good idea,' it's probably too ordinary." That common ground partly explained their immediate affinity for each other.

Dean took Bezos on a tour of DEKA, starting with his favorite place, the machine shop, just down from his office. Like an old-time factory owner, Dean liked to stay close to the "floor," where things were made. He loved to explain what each machine did. Bezos seemed fascinated, too, and asked lots of knowledgeable questions that reflected his long-ago studies in physics, electrical engineering, and computer science at Princeton.

Ginger's marketers—Mike Ferry, Tobe Cohen, and Matt Samson—were waiting for Bezos in the cafeteria, looking eager. Dean had instructed them to grill Bezos. They did, and his linguine grew cold as he offered frank advice.

He thought Ginger would spread quickly through the public, accelerated by word of mouth. "People won't show this to two people," said Bezos. "They'll show it to a hundred people. It will be *viral*. The first question will be 'What is that?' The second question will be 'How much is it?' And the third question will be 'Where do I get one?'"

Director of marketing Mike Ferry mentioned Dean's idea to give expensive special-edition Gingers to fifty high rollers and to auction fifty more, a plan referred to around Ginger as the Bugatti Strategy. "I disagree," said Bezos genially. "It's not democratic. It's not the Model T. You need to sell a lot—soon."

Bezos said he wouldn't do any advertising at all because they wouldn't be able to keep up with the demand anyway. Mike, Tobe, and Matt asked about guerilla marketing and marketing to colleges. Bezos told them not to bother, because it was all going to happen by itself. "In the first year," he said, "the more you spend on marketing, the worse you'll be." Dean jumped up and cupped his ear, wearing an expression that said, "I heard something interesting—my echo."

Bezos recalled the first people who gave Amazon their credit card numbers in 1995, a group referred to by marketers as "early adopters." "They're *crazy!*" said Bezos. He said Ginger would attract the same group.

Dean asked what Bezos would charge. He thought for a moment and named a figure of a few thousand dollars. Dean and the marketers laughed; their target price was almost exactly the same. If the machine debuted at a higher price, Bezos warned, the press would be much different. Since cash flow was always a big problem for start-ups, Bezos suggested asking people to pay for their machines six months before launch, with the promise that the first machines would go to the people who had been waiting in line. Voilà, operating cash.

Tobe Cohen asked how Bezos would go about selling Ginger on the Internet. "I'd call Amazon," he said, laughing that laugh. (The next day one of the cafeteria employees asked me, "Who *was* that noisy guy here yesterday?") Tobe said Amazon was his first choice for an Internet partner.

"You really know how to sit across a bargaining table," said Dean, taking off his watch and handing it to Bezos.

"You need a great name," said Bezos. "I like Ginger." Doug Field pounded the table and said "Yesss!" Dean looked amused and pained. He was still keen on Flywheel.

Given Amazon's fast growth, Doug asked how Bezos preserved the company's culture and managed recruitment. The initial team set the culture, said Bezos, but when the business took off, there came a point when it was tempting "to fill holes with bodies." Resist that impulse, he said, because A's hire A's, but B's hire C's, and C's hire D's. Since the plan called for Ginger to expand within a year, from thirty people to a hundred and fifty, Bezos advised hiring a full-time recruiter. He looked up and down the table. "I think you're in for the rocket ride of your lives," he said.

Someone asked if they eventually would need a human resources department. That hit one of Dean's buttons. "HR is a solution in search of a problem," he said. "I read the policy on sick days from Johnson & Johnson and Baxter, and they're eight or nine pages long, with all these regulations—two sick days before using vacation days, days you can or can't carry into next year. Well, we have a policy: If you're sick, don't come to work; if you're well, come to work. In their plan, if you're sick for more than a few days, the money stops. But your bills don't. Right when people need the money most, they cut it off. We have *never* kept a cent from anyone who was too sick to work."

An HR person once told Dean that people would take advantage of his system. "So since somebody might lie and cheat," he told the lunch table, "we should punish people who really get sick?" That reminded him of a government bureaucrat who told him he had to have a written personnel policy. "So I took out a business card and on the back I wrote, 'Use good judgment,' and said, 'That's my written policy.' "

Bezos honked. "I love to hear Dean talk about bureaucracy," he said.

Lunch broke up. After showing Bezos the iBOT and the Stirling, Dean took him to a big empty space where the Amazon king could ride Ginger. He tore around the room, laughing raucously and blurting, "*Yeah,* baby!" and "It's a *blast!*" As he rode, Dean's assistant called to say that Doerr was on the phone and wanted to talk to Dean and Bezos—separately. "We don't have to do what he wants," said Bezos, grinning.

"I'm afraid this isn't going to be good," said Dean.

On the way back to Dean's office, Dean explained that Doerr wanted to change the deal to give the investors the right to take the company public four years from now. "So I said, 'You only own 4 or 5 percent but you want the right to tell me what to do with *my* company? That's not right, and besides I don't like people telling me what to do.'"

The other thing Doerr wanted was participating preferred stock, which gives investors their money back, plus a rate of return, before anybody else gets a dime, *and* an additional sum if the dividend on the common stock exceeds a certain amount. Bezos said that VCs always asked for participating preferred, but advised Dean not to give it to Doerr; *he* hadn't. Dean wasn't sure what to do, because Doerr called these new stipulations deal-breakers.

"Well," said Bezos, grinning again, "John knows that you have others waiting to invest. Like me."

Moments after we reached Dean's office, Doerr was on the line. Dean put him on speakerphone and explained that Bezos had invited himself to DEKA on the spur of the moment. Doerr didn't seem ruffled by it. They got down to business. Dean said he didn't want the company to go public right away. "I want to build the best product and the best company to change the world."

Doerr agreed with those goals, but nevertheless wanted the IPO clause. He asked Bezos what he thought of Ginger. "I *love* it," said Bezos. "I'm on the Steve Jobs plan. I mean, if I'm wanted."

"So, Jeff, are you going to join the board?" asked Doerr, presumptuous as always.

"We haven't had that discussion yet," said Dean.

The conversation wasn't settling anything. As soon as Doerr hung up, Dean and Bezos walked down to Bob Tuttle's office. Tuttle had been spending a lot of time on the phone with Doerr and Michael Schmertzler, who was fed up with Doerr's dramatics and hotdogging. "John is being John," said Tuttle in his quiet monotone. Dean asked him where the deal stood now.

"We have accepted participating preferred as a concession to John," said Tuttle.

"We have?" asked Dean.

"Yes."

"And you're OK with it?"

"Yes."

"OK," said Dean.

As for the right to force an IPO, Doerr had pointed out that Kleiner Perkins hadn't ever exercised it on a company. "Then why do they need it?" asked Dean.

"Leverage," said Tuttle. Doerr could have that too.

The other sticking point was voting rights on the board. Schmertzler insisted that as the initial investor, CSFB should have two votes. Doerr insisted on parity. The two men were scheduled to talk today, to see if they could work things out.

"Not everybody has to be happy, Dean," said Bezos. "If John can't sign these terms, so be it, and you'll find someone else. I like John, but this is business."

Tuttle asked Bezos if Ginger met his expectations. "Ginger is great," he said, "but I just love DEKA, the informality of it. People are doing real work and the furniture doesn't match."

"Well, when your whole accounting system is your mother and a checkbook . . ." said Dean, shrugging. He himself hadn't written a check in twenty years. He used plastic and handed the receipts to his assistant, who handed them to Mom. He was determined to be a guy in denim and to keep DEKA a down-to-earth crucible of invention where his mother did the books and fussed at people who submitted heavy expenses. But he also wanted to change the world and build a multi-billion-dollar international company and be pursued by heavyweights like John Doerr and Steve Jobs and Jeff Bezos.

This tension was manifest in his thirty-eight-acre estate—he would hate to hear it called that—which he had named Westwind. That evening Dean took Bezos on the tour. Dean liked to show off the house and its eccentricities. It was far more revealing than Dean seemed to understand, a monument to engineering, ego, conspicuous consumption, and the Old Economy. It looked like an impersonal show house, with never a spill on a counter or an item out of place, yet everything about it was deeply personal. Dean told Bezos it had taken him seven years to design every detail of Westwind, right down to the locations of the smallest shelves, and two and a half years to build it.

The 32,000-square-foot hilltop house offered mountain views in every direction. A heated hangar housed his two helicopters. In the garage, a Porsche 928 sat next to a black Humvee. DEKA employees sometimes used the tennis court, baseball diamond, and basketball court, but Dean never did. (At one time he had played his own version of tennis, disregarding the silly boundary lines, but now, aside from climbing stairs while making his rounds at DEKA, his exercise was mental.)

The house was hexagonal, with many semi-open levels that zigzagged through the central space, partially blocking every perspective and creating a maze-like effect that reflected its designer. Even though Dean had shared Westwind for more than six years with K. C. Connors, it was a bachelor's house, with big wooden beams, black steel railings, and leather furniture. It felt more like a severe lodge in the off-season than a home.

The artwork was dominated by paintings and cartoons done by Dean's father Jack, a talented illustrator who had worked for lurid comics such as *Creepshow, Tales from the Crypt,* and *Weird Science* early in his career, before turning to commercial illustration. The rest of the décor consisted of machines that Dean had built or refurbished into gleaming working order: elaborate pendulum clocks, antique drill presses, little steam engines, a small fan with a Stirling engine, the diesel that once powered Henry Ford's yacht, and something he called "a one-lung fire-breather." "I like the physics they represent, the craftsmanship, the sensuality," he once said. "I like the *sounds* they make. And they are moving, breathing pieces of history."

Bezos lingered over the machines, examining them and asking questions. "I think we've forgotten how to build stuff," said this monarch of the New Economy.

Dean, devotee of the Old Economy, nodded vigorously. "The world is *about* cool stuff."

Some of the machines were Christmas gifts from the engineers at DEKA. Dean explained one of his favorites to Bezos—the south-pointing chariot. The Chinese had invented the device more than two thousand years ago for military purposes. Dean demonstrated: No matter which way he turned the wooden chariot, the arm on top always pointed in the same direction—south—thanks to a complex system of differential gears. The mechanism fascinated Dean not only because of its ingenuity but because the Chinese had discovered the magnetic properties of lodestone around the same time, yet for several hundred years they had treated the compass as a magical toy. Dean's engineers had built the chariot because he cited it so often. Was it a brilliant invention or the first absurdly complicated concoction by a military contractor? It became DEKA shorthand to ask, "Is this a great idea or a south-pointing chariot? Will it soon be outmoded by a simpler, better idea?"

To end his presentation to Bezos, Dean pressed a button on the chariot, causing a figurine to pop up—Einstein with a compass in his belly. Big honking laugh.

Just down the steps from the main living room, Dean maintained a full machine shop, with several lathes and grinders, including a massive Bridgeport. In the sub-basement, he proudly presented the electrical system that ran everything—3,000 amps, with three-phase power (single-phase is standard). Down the hall, the climate-controlled wine cellar was lined with a couple thousand bottles chosen by a friend who had tried to teach Dean about wines. "I don't have enough mental space for that," he said. "I need the space for other things." He had his own vanity vintage, Westwind, with a label featuring a line drawing of the house. He and I once drank a bottle of it in the wine cellar. It was the first time he had ever sat in this gorgeous hexagonal room and done that, though he had been living in the house for several years.

Tonight's tour skimmed past the game room off of the first floor, with its pinball and slot machines, its Ping-Pong and pool tables—none of which Dean ever played—and its Wurlitzer jukebox stocked with oldies and classical music. On the bottom floor, Bezos chuckled at the whimsical indoor pool room, its sculpted rock grotto with twisty passages that led to the sauna and exercise rooms. Dean never used any of it.

For a chef, the spacious kitchen was a dream, with yards of marble countertop running between sinks and a commercial stove. There were several refrigerators, one of them devoted to beer, bottled water, and green tea. Dean never cooked and rarely ate at home. The dining area resembled a conference room, with an enormous table and more than a dozen leather swivel chairs.

In his office library, where he spent most of his time at home, he demonstrated that if he tilted a certain book (*Ingenious Mechanisms for Designers and Inventors*), the wall opened onto a secret staircase that led to his bedroom. Bezos liked that. At the peak of the house was a glass observation tower, which appeared on aviation maps as the highest point between Bedford and Boston. To reach the tower you passed through an inconspicuous opening on the flank of a massive stone chimney in the living room, then climbed a concealed winding staircase for two stories.

"You know how when you're a kid," Dean said, "and you'd like a house with all kinds of cool stuff, and then you grow up and you decide you really can't have the fireman's pole and the secret passageway through the bookcase? Well, I decided I *would* have all those things, and I do."

The next day Dean and Bezos met Michael Schmertzler in Dean's conference room. They stood around the table for an hour and a half, mostly talking about John Doerr. Schmertzler had just gotten off the phone with him, without settling anything, which led him to speak with uncharacteristic bluntness. "John's teeth only go in one direction," said Schmertzler, mimicking a shark's bite by curving his fingers backward near his mouth. He told Dean and Bezos that he had been "spending his capital" with his boss, Jack Hennessy, who'd had it with Doerr's shenanigans.

Bezos remarked that his experience with Doerr had been similar. But five minutes after he had told Doerr to walk, Doerr called back to OK the deal. Dean sympathized with Schmertzler. Schmertzler had bent over backward to accommodate Doerr, yet Doerr was acting as if *he* was the gracious one for accepting equality with Credit Suisse. "It's some East Coast/West Coast thing in his mind," said Dean. And maybe some presidential politics. Hennessy's impassioned support for Bush was matched by Doerr's for Gore.

Bezos wondered why Kleiner Perkins was treating Ginger like one of its typical "deli deals" made over sandwiches with a couple of twenty-somethings peddling a concept. Dean mentioned, in a dumbfounded tone, that Kleiner Perkins had asked for the right to replace him if they weren't satisfied with his performance. Bezos and Schmertzler said that venture capitalists commonly demanded that power. Dean shook his head. "I could *never* sign that."

Afterwards, Dean walked Bezos to the FIRST building across the street, where he intended to pick some Amazonian fruit. On the way, Dean said he understood Doerr because they resembled each other. He worried that Doerr might feel bitter toward him if Kleiner Perkins ended up backing out. For all his pugnacity, Dean still wanted everyone to like him, especially his investors.

"You can't really worry about that," said Bezos. "It's just business."

Dean hoped that his investors would say the same thing when they heard that Ginger's schedule had slipped nine months. The first board meeting was just around the corner.

chapter ten

No More Shakespeare

July 2000

After the push to get D1 out, Doug had asked the team members to limit themselves to eight hours per day for a while so they could recover a measure of personal life. Doug ignored his own advice, of course. J. D. Heinzmann tried to comply, but it wasn't easy. With his wife in California, he had lots of time and no outside obligations, and he loved working on Ginger. But he forced himself to redirect energy into other projects, such as rehabbing his Victorian house. He also put in time on his 16-foot dinghy, built in 1915, whose lapstrake construction was hell to refinish. And he resumed tussling with his 1962 Datsun Fair Lady, a rare sports car. He couldn't get the rear wheels off, a dilemma he described as "a real lesson in axle design."

One day in July, the engineers were joking about a news story concerning a Russian engineer who had invented boots that fired a piston on the downstep, propelling the wearer up to 13 feet. That reminded J. D. of his old Hop-Rod, a boyhood gift from his father. He brought it in the next day to show everybody. It was a pogo stick with a small gas engine attached to the spring, which in turn was connected to C batteries in the shaft. You jumped on it once or twice to spark the motor, which then drove the stick up and down with a pop-pop noise. The day he got it, J. D. had done a thousand hops without stopping. The next day he could barely walk. When it was suggested that the device was crazy, J. D. agreed, for the wrong reasons. "I would design it with a longer stroke, to make it easier on the calves. They banned these for some reason," he added, sincerely puzzled.

J. D. and I left the Hop-Rod and went to lunch with his brother Kurt. They were fresh from a July 4 family gathering. One night they had stayed up late smoking cigars and drinking gin-and-tonics. At 11:30 their younger brother, who had returned to college to study chemistry and engineering, said, "Let's light some magnesium." Being a Heinzmann, he happened to have a bar of it handy. Being Heinzmanns, his brothers thought that lighting some was a grand idea. Evidently a common one, too, since their brother had waited to suggest it until after their parents had gone to bed. "Because they hate it when he does that," explained J. D.

So they cut off a hunk of magnesium with a band saw and drilled holes in it for easier ignition. The thrill was that magnesium burns intensely white, and once it's lit, you can't douse it. "Water only makes it brighter," explained Kurt, "because it breaks the H_2O molecule and burns the oxygen."

The episode reminded J. D. of how Kurt, as a teenager, had discovered that if you put propane in a baggie and then stick in the lit tip of a propane torch, you get a big bang. Kurt hatched the idea after noticing that the torch popped when you lit it, so logically if you increased the area, you should get a louder pop. He experimented with bags of different sizes. "But a baggie was as loud as you wanted it," he said, meaning it was incredibly loud. "And it doesn't just go *bang!*" he said. "It goes *BANG!-ANG!-Ang!-ang!* It makes your head pound."

"And he did it the first time indoors," said J. D., "in my parents' house." The brothers looked at each other and grinned.

Dean rarely ate breakfast or lunch—too bothersome—but one day in early July he was having a sandwich in the cafeteria with Bob Tuttle and talking about the upcoming board meeting, the first ever, in which Ginger's investors would hear that the schedule had slipped nearly nine months. Dean blamed Doug. He scoffed at Doug's justification, lack of personnel. What about the handful of contract engineers he had been using? When Dean was displeased, he often used Tuttle as his messenger.

"Try the fatherly approach with Doug," said Dean. "He's very sensitive and very enthusiastic and I don't want to harm that. But he's got to understand that a board meeting is not a love fest. We don't need Shakespeare quotes. People want to know *when* it will be ready and *how much* it will cost. The schedule is now radically different, but he won't admit he made a mistake. He needs to do that at the board meeting."

The original launch date of May 2001 had been set in April 1999 by Dean, Bob Tuttle, Tim Adams, and Doug. Tim considered the date a "thumbnail" guess, so its inaccuracy didn't surprise him. By the first quarter of 2000, Doug and Don Manvel both had warned Tim that May 2001 wasn't possible, and Tim had told Dean. So the principal players had known about the slip long before Doug led his engineers to the realization in June with that backward chart. But May 2001 was the launch date that Dean had used with investors, so backing off it by eight or nine months looked bad, and Dean wanted someone else to take the fall. Never mind that he had agreed with the original blue-sky estimate.

Dean's real beef was that none of the managers had realized just how far off schedule they'd gotten. Tim felt partly responsible. He certainly should have foreseen the severity of the miscalculation. But he was sure that Doug didn't have enough resources, and he disagreed with Dean that the slip was mostly Doug's fault. Even if Doug's team had been able to meet the original deadlines, Don Manvel also was behind, because of Dean's demands about easy terms from suppliers and his reluctance to approve a manufacturing site. The funding had come in late, too, because of the intrigues between Dean and the investors. Tim told Dean that no one person or program was to blame, which Dean took to mean that Tim was avoiding responsibility.

In the days before the board meeting, the team members rehearsed their presentations. Tuttle told Scott Frock, the new financial manager, not to say "cash burn," but "spending to launch." After listening to Doug's rehearsal, Tim said, "I'd add another dot: 'We underestimated the task.' There's nothing wrong with that and it's the truth. If you say it was all because the money came in slow, you're opening yourself up for attack. You put together the plan fourteen or fifteen months ago, and there's nothing wrong with saying you're smarter now."

Tim and Dean considered giving the board the choice of an earlier launch, starring either the urban Metro version of Ginger or the all-terrain Pro, but not both. They decided against that idea. If they proposed an earlier date and didn't hit it, they would look bad twice. On the whiteboard in the War Room, Doug had written, "We are there later but we are in a better place. Stronger, Smarter, Better Protected." This, in fact, was how Dean proposed to spin things for the board: The new schedule would give them more time to reach their targets for cost and volume supply, more time to overcome regulatory problems, more time for consumer research and launch strategy. All of this would improve their chances of long-term success.

The board members arrived in Manchester the night before the meeting and ate together. They schmoozed and rode D1. Dean reminded them that the idea was to change the world, not to score a quick financial bonanza. When they were all feeling warm about the machine and the future, Dean mentioned that tomorrow would bring some bad news: a big slip.

The next morning at 8:00, Tim Adams, Bob Tuttle, John Doerr, Michael Schmertzler, Paul Allaire, and Vern Loucks gathered in Dean's conference room. Dean still hadn't decided what to do about Jeff Bezos and Steve Jobs. After a few pleasant sentences about the exciting trip ahead of them, Dean said, "As you know, I've been very determined to keep control of the company. If I've been a hard-ass about it," he continued, "that's the way it is. You're investing in *my* company. But I also recognize that for some period of time it's all *your* money, and it's my job to make as much money as I can for you. If reasonable businesspeople think that taking the company public is the best course, we'll do that."

"The technology is so good," he added, "that the risk is not that it won't be everywhere in ten years, but that it might not be *us* providing it—it'll be Honda or Sony." He asked them not to debate things today, but rather to soak up all the information that they would be hearing for the first time. He also asked them not to beat up the team. "This may sound hokey," he said, "but some of them are very sensitive. So if you want to beat up on someone, beat up on me, because I'm responsible for your money." He assured them that the team understood the seriousness of the slip, and he put the rehearsed spin on it.

CEO Tim Adams outlined the basic strategy for manufacturing and service, then offered three explanations for the slip: staffing delays caused by cash conservation, the delay in financing, and underestimation of the task.

Dean introduced Doug by saying that his new son had been born three days ago, on July 4. In keeping with DEKA tradition, Dean presented Doug with inscribed silver calipers in a wooden box. "I meant to wrap this," said Dean, "but where Doug and I continually disagree is on the importance of how things look."

Wearing khakis instead of his habitual jeans, Doug presented his information quickly and smoothly. He explained Ginger's components and passed them around. He also talked about giving the machine a soul, and he quoted Picasso and Shakespeare. He didn't apologize for the new schedule.

The board wanted to know which regulatory organizations might rule on Ginger. Doug ticked them off: the Consumer Product Safety Commission, the Federal Communications Commission, Underwriters Laboratories Inc., the Department of Transportation, and all their international counterparts.

"The engineering issues have been considered," said Dean, "but the sixty-four-*billion*-dollar question is whether a cop will stop you. So we're collecting information state by state."

Dean had worried beforehand that operations manager Don Manvel might be mistaken by Doerr and Schmertzler as a Rust Belt plodder because of his age and low-key, halting delivery. In fact, Don was ingenious at convincing suppliers to accept Dean's terms. When Don finished his thorough report, Doerr asked how many machines would be in stock on the new launch date. Doerr was short, slight, and rumpled, with tousled hair, a nerdish combination that made his imposing bass voice disconcerting.

"A better question," said Dean, "would be, what's the run rate?"

"No," said Doerr, "I'd like to ask how many we could *ship* on the launch date."

Don hedged, saying that January '02 was the scheduled start of production.

"So it's not a good assumption that we'll have *any* machines to ship," said Doerr, sounding unhappy.

Scott Frock, the new financial manager, reported that the burn rate at present was $2.5 million per quarter, but that would change quickly as they hired people and geared up for production.

Mike Ferry presented the marketing strategy for launch. He told them that a new product typically needed six years to take off. A few had gotten airborne faster—the cordless phone in three years, the CD player in two years, the camcorder in one year. Naturally he was hoping that Ginger resembled the camcorder. The launch plan called for two versions of Ginger in the U.S. consumer market, a small Metro model for urban use and a bigger Pro model aimed at recreation. In the second year, Ginger would penetrate the commercial and international markets. Several things could imperil the launch: if Ginger was priced too high, outlawed on sidewalks, or perceived as unsafe. Mike mentioned Dean's Bugatti strategy of selling special-edition Gingers to celebrities; no one liked its taint of elitism.

Aileen Lee, a young associate partner at Kleiner Perkins who was Doerr's assistant, asked the marketers to keep in mind that men and women would view the machine differently. Men would like to know how it worked and what was inside, whereas women would be more interested in practicalities such as a place to carry groceries. Schmertzler nodded. His wife had said the same thing.

Someone asked when the marketing team would start consumer testing. Mike Ferry had posed the same question to Dean before the meeting, and now repeated Dean's response: Because of the slip, that date had moved from next month to some vague future date when they would start testing on DEKA spouses. Such data wouldn't be representative, Mike said, but it would be better than nothing. Which, he noted dryly, was what they had now. Mike didn't tell the board that he felt frustrated by the new delay and shackled by Dean's insistence on secrecy. How could he do his job if he couldn't test consumers' reactions to the machine?

Vern Loucks wondered aloud about the wisdom of putting off the testing. Dean jumped in. The question was when to wake sleeping giants. "Companies like GM and Honda spend billions on R&D," he said, "and they'd spend a billion on this without blinking. The only advantage we have is that they're clueless."

Secrecy was a thorny issue because it impinged not only on consumer testing and competitive advantage, but also on the central issue of getting Ginger onto sidewalks. Regulators worried the board. They tossed the issue around. Allaire and Loucks, big, thoughtful men, spoke in pithy sentences. Schmertzler asked his usual crisp questions. Doerr sprayed bursts of words. Allaire couldn't imagine regulators letting Ginger cruise at 15 miles per hour among pedestrians. Doerr agreed. Schmertzler noted that Ginger had grown from a slim little prototype to the pumped-up D1, which he thought too big for the sidewalks of New York.

So when should they risk disclosure by showing Ginger to city leaders and regulators? Doerr suggested six months before launch, but Allaire thought that was far too late. Regulators were very deliberate, he said, and if they felt rushed they would take the easy route—prohibition or excessive regulation. Dean disagreed. "In my experience, you never ask a bureaucrat for permission. There's something to be said for letting it build until the bureaucrats can't say no." Schmertzler cut to the point: When it came to the regulatory bureaucracy, none of them knew what they were talking about. The solution was to hire an expert as soon as possible.

"Right," said Doerr. "Next topic."

Funding. Kleiner Perkins was investing $38 million in two installments, and Doerr was still being evasive about turning over the second chunk, still trying to gain more edge on the deal. He had been driving Michael Schmertzler mad and had even drawn traces of exasperation from silicon-based Bob Tuttle. Today, Schmertzler and Tuttle had investment allies who didn't like to be teased.

"When do we close the door?" asked Loucks in his gruff way.

"In the next couple of days," said Tuttle. "Right, John?"

"Right," said Doerr unenthusiastically, without looking at anyone.

"It will either be settled or not," added Tuttle.

"Good," said Loucks. "Nothing against anyone here."

"I'd like to agree with Vern," said Allaire. "We need to get it done within a week. You either do it or you don't."

Doerr stayed silent, but he got the message: Put in your chips or get out of the game and let someone else play. (A week later an attorney from Kleiner Perkins called Tuttle and settled the funding.)

They began discussing the hiring of a chief financial officer and a chief information officer. Tim Adams said it was hard to find people they could afford. Doerr came back to life, looking aghast. "That you can *afford*?" He gave Tim a brief lecture about how crucial these positions were to Ginger's success. Tim knew that. The problem was Dean's parsimony, but Doerr wasn't aware of that, so Tim looked bad. Dean didn't defend him. Dean was willing to protect his team by taking the board's heat for the delay, but not to defend his CEO against criticisms that stemmed from his own deficiencies as a manager—deficiencies that the board still hadn't recognized.

The meeting broke up in the early afternoon. The board had heeded Dean's wishes and not attacked the team about the slip. They hadn't attacked him either. In the hall Dean said to me, "That couldn't have gone any better. Of course, it's not over yet."

Dean left to ferry Allaire and Loucks to the airport in his helicopter. Both men told him they were pleased with Ginger, but there had better not be any more big mistakes. Allaire said, "Does Tim know that if he does this again, he's out?" Dean marveled at Allaire's decisiveness about personnel. When an early director of FIRST fouled up, Dean had asked Allaire, then FIRST's chairman, what to do. "What do you mean?" Allaire had said. "You fire him." Dean hated to fire people. He would agonize about it for weeks.

By the next day, Doerr and Schmertzler also had complained to Dean about the slip. No surprise there. But it worsened Dean's irritation that Doug had estimated so poorly, and that Tim hadn't known it was going to happen.

"Doug, who has charts up the gnat's ass," said Dean, "and he's surprised that he lost nine months? How can these guys be so intelligent and let this happen? Like, 'Mrs. Lincoln, I know you're upset, but the show was so good.' But I think Doug thinks, 'I did the best I could, I made a lot of charts, so now I'll just make some new charts.' If it was anybody but Doug, you'd kill him."

I suggested that Doug might have been focusing on the D1 deadline. "But he was swatting flies while he was getting trampled by the elephant," said Dean. "The goal in life isn't to make people feel good. I want them to feel good *after* it's done. I just wish they'd spend a little more time on engineering and less on poetry." He also wanted Doug to be the one who scolded his team. "I'm tired of being the guy who shoves spinach up everybody's nose."

"I've been sitting here for hours trying to figure it out," he continued. "Part of me is happy that the board meeting was anticlimactic.

I'm feeling good that we got the money, we've got this world-class board, and they want to do the long-term thing that will change the world, not some quick turnaround for cash. But I'm also worried about the cliff we're heading for. The nagging thought is, how do I get my own guys to understand that we never let this happen again?"

When Doug Field first realized how big the slip would be, he panicked. But panic was an emotion he couldn't afford, so he simply stopped thinking about the slip and concentrated on D1. By the time that was finished, he had become philosophical about the slip. He felt severely understaffed and refused the role of fall guy. Shit happens. Deal with it and keep moving. But he knew that the slip put him under even more pressure. The first guesstimate of the schedule had been hazy. The new deadline was a clear, deep line in the dirt.

To get there Doug needed more engineers, especially ones with expertise in embedded systems, electrical integrity, and mechanical design and release. Only the best would do. The most tempting pool of candidates was DEKA itself, but Dean had put up No Fishing signs there months ago. Nevertheless, DEKA engineers kept banging on Ginger's door, the portal to Candy Land. At the Christmas party, someone had made a poster that pictured Fred in a dungeon and Ginger among some dancing gumdrops. An engineer from Fred once said to Doug, "If Fred is hell and Ginger is heaven, you must be Jesus Christ. I hope you'll save me."

This attitude irritated Dean. "Who wouldn't rather make ice cream than broccoli?" he said. "But somebody has to keep working at Fred." People at Fred resented the airs of Ginger people, he said, illustrating it twice in the same conversation with a disdainful mimicry: "At Ginger we do it *this* way." In other words, not the DEKA way— not Dean's way. Dean hated the idea that the Ginger team considered itself different and separate from DEKA. He disliked this child's growing independence.

But independence was what the Ginger team longed for. They preferred Doug's and Tim's management styles to Dean's. "Dean cares about his relationships," said one engineer at Ginger, "but he'll also tell you he doesn't care what anybody thinks about him. He'll do things his way and say what he thinks, and to hell with anybody who doesn't like it. Doug is the antithesis of that in leadership. I wish Dean could see the value of it and acknowledge it."

In fact, Dean prized Doug and recognized the cohesiveness he had inspired. But as Ginger moved closer to production, the team began to diverge more and more from the DEKA model in methods, goals, and perspectives. Dean seemed to understand the necessity of this—unlike other engineers at DEKA, the Ginger team had to finish a machine for mass production—but he also resented it. So he mocked Doug's charts and stickies and the team's "airs." By July it was clear that Dean wasn't going to let the team move out of easy reach, beyond the Millyard. As a temporary fix, Don Manvel's team was erecting a "plywood plant" downstairs, to build some test Gingers.

Meanwhile, the team leaders kept trying to recruit people, especially engineers, which meant spending precious time to screen candidates and do interviews. They needed a hiring manager, as Jeff Bezos had suggested, but Dean wouldn't allow that extravagance. Though the market for engineers was extremely tight, he continued to insist on low-balling salaries. At a meeting about recruitment, one of the team's managers joked that he took whatever rock-bottom salary he thought might attract a candidate and then multiplied it by 0.8—"the Dean Factor."

"0.8?" said another. "Are you doing that well?"

Everyone laughed, but the truth inside the joke wasn't funny. The project existed because of Dean's ingenuity, tenacity, and salesmanship, which were also elements of the Dean Factor. But Dean often threatened Ginger's health by undermining its managers and then blaming them for problems. The Dean Factor was sometimes a multiplier, sometimes a divider.

The week after the board meeting, Mike Ferry, Tobe Cohen, and Matt Sampson met with Lexicon again about Ginger's name. The three marketers left ecstatic about a new suggestion: Landwing. Tim loved it and so did Doug. The marketers presented Landwing to Dean with great excitement, but he pronounced the word as if he had a speech impediment—no, as if it *caused* him to have a speech impediment. So Landwing was DOA. Dean wanted Flywheel, even though the marketers now felt that it might connote dangerous speed to regulators.

Sometimes the conversations between Mike, Tobe, and Matt were as packed with idiosyncratic jargon as the engineers'. They discussed how to make the Ginger virus spread faster among consumers, how to do an end run around the usual acceptance curve, how to compress the takeoff point so that diffusion exploded. They wanted to ladder-up quickly and avoid triggering the public's puke-o-meter.

Mike reported to Tobe and Matt that the board seemed worried that the first users of Ginger would be young hotshots who might worsen regulatory problems. The board seemed to prefer a safer launch among older, wealthier consumers. The marketers worried that if Ginger aimed at that demographic instead of cool early adopters, the machine might not stir up the next waves of buyers, the "early recruits" and "draftees," and instead would wallow in the Sharper Image niche.

But they were just shooting arrows into the dark, because they still weren't allowed to test how consumers would react to the machine. So they improvised. They went to New York City with Doug Field, Scott Waters, and Tao Chang to do some experimenting. Using a kick scooter, an electric scooter, and a small folding bike, they cruised sidewalks and stores throughout the city, gauging how people responded to unusual transporters. They rolled through Penn Station and Grand Central. They scooted onto the subway. Whenever they saw a cop or security guard, they made a point of whisking right by. Though a cop in Grand Central yelled at Doug to get off the Zappy, everyone mostly ignored them in good New York fashion. They even scootered past the guard in the glassware section of Williams-Sonoma without raising an objection.

They also tested a box on wheels with the same dimensions as Ginger, pushing it along sidewalks and through Grand Central at rush hour with no problems. But all the stairs and escalators would be obstacles unless Ginger was easy to lift. Nor could they get the model through subway turnstiles. On the sidewalks, they had to maneuver through crowds to reach the curb cuts, as would someone on Ginger. Water tended to accumulate up to six inches deep in the curb cuts, so Ginger's chassis would have to be waterproof. The aisles in New York's grocery stores were very narrow, but lots of people used metal mesh carts, 24 inches wide and deep, with no problem. The city's many metal grates and plates would be slippery when wet. The engineers needed to consider all these things.

Doug unrolled a long chart in the War Room. It ran the length of the table and was covered with stickies. Dean stared at it. It illustrated the schedule between July 2000 and launch in January 2002. Doug began explaining the various phases. When Dean asked why the Metro build occurred so late, Doug picked up that sticky note and placed it a couple of months earlier. ("That's what worries me," Dean said later. "Having problems? Just change the chart instead of doing what needs to be done. Why don't they just build one in a couple of days and see what it's like, instead of charting everything?")

Still studying the time line, Dean asked Don Manvel how many V1 machines they could build in August 2001. V1 was the last rev before production began.

"As many as you want. If you want five thousand, we could do it."

"I might want five thousand," said Dean. "We call them Betas and we charge $10,000."

"But we'd need to support those with service," said Don.

"Not really," said Dean. "Our service is to send it back. I'd bet your burn rate by September 2001 will be $3 million a month. Wouldn't it be nice to have $15 million or so coming in?"

He waved his hand dismissively over Doug's chart. "I have to turn this into something that fits into *my* head," he said.

In late July Dean hosted an evening meeting at his house for potential investors in the Stirling engine project. Bill Sahlman, the procurer of angels, showed up first. We stood on the deck watching the sunset melt over the hills. Dean suddenly brushed his head. "The bugs are coming out," he said. "Let's go in." He hated insects. If one somehow got inside his castle, he left the room or called K. C. to come kill it.

The rest of the group arrived and sat around the big wooden table in Dean's dining room. John Doerr and his associate Aileen Lee were there. Michael Schmertzler was cruising the Greek islands, but he sent two representatives from Credit Suisse.

Doerr finally had consented to be equal partners with Credit Suisse on the Stirling. Since the Ginger board meeting, he and Schmertzler had become allies. For Dean, that solved one problem but created another. The two financiers were working on a joint proposal about the

Stirling funding. "I have both foxes deciding how to divide the hen-house," Dean said. Schmertzler thought Dean was being cynical. "OK," said Dean, "then you won't mind if we get a proposal from Sahlman's group as well."

In his welcoming remarks, Dean acknowledged Doerr's recent generous contribution to FIRST—$1 million—and the two-page story in that week's *Time* magazine about his philanthropic work. Then he started selling them on the Stirling. As always, it was a masterful performance, despite occasional bogs of techno-speak. He noted that businesses paid huge fees for access to emergency servers in case of a power failure, in order to save computer data. But businesses could buy his Stirling engine relatively cheaply and run it for pennies as an emergency backup. It was small enough to sit on a shelf in a closet, with no need for venting.

He told them that a competitor was working on a Stirling with several elegant features, including a linear alternator. He grabbed a napkin and began to draw and explain, talking excitedly about current and magnets and dissipation and square roots, all of which led him to conclude that the competitor's engine was doomed to produce unreliable power because the design flouted physics. "Not the first law of thermodynamics, the second one, the *mean* one."

His listeners didn't know what he was talking about, but he plowed on, enthusiastically explaining how a competitor made a Stirling engine with no bearings—another elegant but, alas, flawed design. When he finally stopped the technicalities, Doerr asked how quickly Dean could get his Stirling to market.

"I anticipated you would ask this question," Dean said dryly, raising chuckles around the table. "I don't know, but I'll give you data. It would probably take a year to tool up."

"A *year*?" said Doerr.

"Yes," said Dean. "That heater head is really unusual. Do you really want to spend $30 million tooling up on the two we've made, or should we build another twelve and beat the hell out of them, run them for a couple of months? So you do some number of months for design, some number of months for tweaking, and then you put a stake in the ground and you tool."

"But *why* does it take a year to tool?" asked Doerr. "That's the question on the table."

For dot-com people like Doerr, Dean once cracked, every week is a year. They didn't understand that products were different from bandwidth. So Dean told Doerr about Boeing, where it took three years to tool up for a new airplane, and about Harley-Davidson, which took the same amount of time to tool for a new motorcycle. "So I'm just telling you that it takes a *long time*." Doerr was slumped with his head in his hands. Sounding resigned, he said, "So it's a year on design and iterations, and a year to tool?"

"Plus three months for testing," said Bob Tuttle. Doerr slumped a little more.

Dean noted that a lot of companies were trying to make new energy devices with Stirling engines or fuel cells. That led him into a digression about lithium batteries and their inefficiency, which led him to cite Euclid's formula for calculating the volume of a sphere, $4/3(\pi)r^3$, and gamma defined as "a measure of the nonlinearity of specific heat." His listeners were getting glassy-eyed again. Dean wanted to talk about engineering, but these potential investors were looking for different information.

"How would we build a team?" asked Doerr. "That's the reason I wanted you to meet Norbert"—Norbert Gottenburg, a Kleiner Perkins partner who specialized in high-level recruitment, was sitting at the table—"because I'm thinking you'd want a world-class CEO within a year." Michael Schmertzler might have been a little wary about this offer to let Kleiner Perkins handpick Stirling's CEO.

"I wouldn't think of adding a guy with a tie who goes around the world," Dean said. "I'd add a bunch more technical people first." Once again, the dot-com mind-set collided with the engineering mind-set.

The Stirling team was developing two engines, one the size of a carry-on suitcase, the other slightly larger. The larger one made 2.5 times more power, cost 30 percent more, and was less mature in design. Dean wanted to commercialize the smaller one first. He noted that Schmertzler probably would disagree, because the larger one was a better bet to make a lot of money fast. "But eighty-five thousand people died today because they didn't have good water," said Dean. "So to me the *fallback* position is to sell a hundred thousand larger ones a year to yuppies at $5,000 each."

"I vote for doing both as fast as you can," said Doerr.

Afterwards, when everyone had left, Sahlman, Dean, and I drank some Westwind wine. "Dean," said Sahlman, "you are the single greatest selling machine I've ever seen."

Dean looked pleased. "Because it's something I believe in," he said. Once, talking about how he picked his projects, he had told me, "If you work on something for five years, that's nearly 20 percent of your productive work life. Most people don't think about that, but it's the main thing I think about. I think about it *all the time,* much more than money. Because the only thing we can't make more of is time."

A dozen members of the Ginger team met in the War Room to hash out the machine's basic architectural platform. That platform would constrain, allow, or disallow other options. They needed to decide which features *must* be included, which must *not* be included, and which should be options. Phil LeMay handed out a chart that listed six possible models of Ginger, ranging from "Minimum Ginger" on the left through increasingly elaborate models of the Metro to the fully equipped Pro. Phil had listed the costs of some of the electrical parts required by these models, which increased from $12 for the Minimum Ginger to $59 for the Pro. Multiply that cost difference by millions of Gingers and the reason for this meeting becomes clear.

The attendees included engineers, marketers, Tim Adams, and Don Manvel from operations. On the first go-round, they voted overwhelmingly to make certain features standard. On other matters they disagreed, usually with the engineers on one side and the marketers on the other. To the engineers, beauty meant simplicity and low cost. To the marketers, beauty meant a machine loaded with features.

"I'm wondering if this is a reality-based conversation," said Tobe Cohen at one point, drawing smiles from some engineers, who often wondered what mysterious dimension the marketers inhabited. "Because if we don't have some of these options," continued Tobe, "we don't have a consumer-friendly machine."

They decided to redo the Must list, this time from the standpoint of cost. Most of the features were relatively inexpensive, except power-on security and LCD gauges for visibility in sunlight.

"I could never give up power-on security," said Tobe Cohen.

"But even if you have that," said John Morrell, "someone can still pick it up and carry it away."

"But then it's a paperweight," said Tim Adams.

"Unless you have a chop-shop of MIT students who can pull the key switch and put in another," said John.

"We want the consumer to be able to step off and not go through a long involved process of locking the machine if he wants a cup of coffee or a paper," said Mike Ferry. "There has to be a way to easily secure it."

But theft wasn't the only reason for power-on security, said Phil LeMay. What if the machine was sitting in a garage and a child could step on and start it?

Nevertheless, Doug abruptly put power-on security under Options. "I think it's an insult to charge someone $2,000 and then make them pay extra for security," objected Mike Ferry.

Doug disagreed. He was trying to push the machine back toward the Minimum column. "It's *not* a vehicle," he said. "The Minimum is the power pedestrian." He rapped his knuckles on the whiteboard. "That's really all we need."

The meeting had raised more questions than it had answered. Doug had reminded everyone that the point of Ginger was the joy of riding it, not whiz-bang add-ons. Afterward, several frustrated engineers lingered in the War Room. Phil LeMay summarized the meeting: "It costs too much; everybody wants everything; find a solution."

"But at some point," said Ron Reich, gunning his motor, "we take our best gut feel and *go* with it instead of talking around and around all the options."

"I'm concerned that some of the nonengineers don't understand some of the issues," said John Morrell. "This 'lots of options' idea is scary. We don't want to clutter it with a lot of fluff."

"No one is not going to buy a Ginger in the first year because it doesn't have one of these options," said Doug. "These guys come from very mature industries where the goal is to differentiate yourself from others. We won't have that problem."

That night Dean hosted a party for the Ginger team to celebrate the arrival of nearly $90 million in funding. The Wurlitzer pumped out old rock 'n' roll. Bartenders poured beer and wine. Dinner was salmon, pulled pork, vegetables, and key lime pie. A few people, especially J. D. Heinzmann, took full advantage of the amenities, swimming in the pool and steaming in the sauna.

Phil LeMay was stroking one of the huge flywheels. He wanted to help Dean fix them. "You don't get to work on the big iron very often anymore," he said. Later he and I played pool for a long time—one game of eight-ball. "I understand the physics," he said. "I just can't translate that through a pool cue."

After the party broke up, Dean and I talked in the library. Bezos wanted in for $10 million. Dean didn't want to take money from both him and Jobs because that would unbalance the board with a West Coast triad (including Doerr). He was leaning toward Bezos, but Bob Tuttle preferred Jobs because he had launched innovative hardware.

Meanwhile, Dean was worried about a profile of him coming out in *Wired* magazine. He had told the reporter about a secret project being funded by John Doerr, but the reporter had just left Dean a message: Doerr had denied involvement in any such project. Dean didn't want to call Doerr a liar, but the alternative made *him* a liar. Doerr's denial puzzled him. "When it's clear that he lied," he said, "won't people trust him less and not believe him next time? Isn't that how it works?"

chapter eleven

Sanctioning Bodies

August 2000

fter the July board meeting, Ginger's direction changed. The larger Pro model, epitomized by D1, had been the project's "prime driver," but now the smaller urban version called Metro zoomed to the fore. Dean, the investors, and the Ginger team stopped thinking about how to dazzle young recreational consumers and began considering ways to woo regulators. Ginger needed to be sidewalk-friendly and usable by millions of consumers. Otherwise it would shrivel into a niche.

This shift altered the machine's design from top to bottom and dumped a special burden on Scott Waters. For instance, D1's large oval dashboard, which the engineers called "the football" (it resembled a toy one), needed to shrink drastically. The handlebar itself, currently a W-shaped bar, now looked like a mistake. The wheels, fenders, and control-shaft base all needed rethinking. As he stood at his high table sketching new ideas, Scott felt regulators peering over his shoulder.

"When we designed D1," he said, "we weren't really thinking about sidewalk compatibility, and now that has become the most important thing." He snorted unhappily. "We thought we were being very aggressive in our design and that we wouldn't have to change much later, but now we're changing *everything*." He paused. "I'm shitting in my pants, to tell you the truth."

For Scott and the engineers, the redesign offered the opportunity, or perhaps the trap, to create lots of commonalities between Pro and Metro. A common chassis and control shaft, for instance, would cut tooling costs, simplify assembly, and allow suppliers to stock one part instead of two. But the launch clock was ticking, and it was much easier to design two unique machines than to force commonality. Eventually—and that word skims over much anguish and many tense late nights—the team decided to go with two unique designs and a few common parts.

That decision still left everything to do for the smaller Metro. Scott, usually genial and easygoing, often looked tightly coiled. One day in August after a meeting that explored necessary design changes, he blurted to design engineer Kevin Webber, "We *can't* do this the way we did Pro. I'm worried that industrial design is going to get pushed aside again. Not by you, but maybe by Ron. So I'm starting to get *very angry* and I'm trying to hold it together." He extended his arms, his hands clenched into trembling fists. "We're talking about designing a new machine in two months. The *idea* of that . . . It *can't* be like it was on Pro."

Kevin knew how to soothe Scott. He had done it often. "Right," he said, "because you know the way we did Pro: 'Here's the wheel deadline, so let's back up two weeks for ID. Here's the control-shaft base, so back up two weeks for ID.'"

"I *can't* do it that way again," said Scott, agitated. "I have to have input at the *beginning* of the process." He took some deep breaths, then apologized for "going off."

Kevin later parsed Scott's frustration for me. Ideally, an industrial designer has time to fill a wide funnel with concepts and then work with engineers to narrow things down. But for each piece of D1, the team usually started with an engineering concept driven by a deadline, and then kept modifying it. "So I'd give Scott a drawing," said Kevin, "and he'd tweak it, and I'd be tearing it out of his hands to meet the next deadline, and so on. It drove us both crazy."

Picking one concept early looked like a way to save time, but that was an illusion, according to Kevin and Scott. In the long run, extensive tweaking required more time than searching for the right design up front. Because they needed to design a new Metro quickly, Scott worried that the funnel would get flattened by Ron Reich again.

Delphi Automotive Systems, the automotive electronics giant, was on Don Manvel's short list as a possible supplier of Ginger's electronics. Dean liked this news. Delphi's CEO, J. T. Battenberg, had become a big supporter of FIRST, and Dean knew that Battenberg wanted to position Delphi as an innovative company independent of General Motors. Dean was tempted to show Ginger to Battenberg, hook him, and milk the resources of his $30 billion company. On the other hand, Dean considered Delphi one of the sleeping giants who could crush him.

"The good news is that they could do all of Ginger," he said, "and that's the bad news too. I think we could become J. T. Battenberg's prize customer, and he doesn't know it yet. He's busy trying to sell a few more dollars of electronics to the car companies, versus the overwhelming content of his material in our machine. But he's got to know we can buy it somewhere else."

"I agree," said Tim Adams. "If it got to the point where he was supplying 65 percent of our parts, I'd be terrified."

"Because then he goes from being a supplier to a partner," said Dean, "and from a partner to an owner."

Because of the tight deadline for the new Metro, the team was desperate for help with the redesign of the dashboard. They hoped to persuade an outside company to loan them a squad of experts to co-develop the part. Delphi was on the hit list because of its expertise in electronics, materials, dashboard lighting, and ergonomics. But when Doug Field, Ron Reich, and Bill Arling visited Delphi's engineering department in Flint, Michigan, they came away deeply unimpressed. By 3:30 the place was half-deserted and half-dark. Bill had never worked in the automotive industry and wondered whether Flint was representative. "It's a typical GM environment," said Doug.

"I got the feeling I was right back in there," said Ron, shaking his head as if he'd had a bad flashback.

Bill and Ron wanted to cross Delphi off their list. Doug said that might not be an option. Dean, Tim Adams, and Don Manvel all liked the idea of Delphi as a supply partner. If the design group was the only branch of the Ginger team uncomfortable with Delphi, they might have to live with it.

"Every time we asked them a question," said Bill, "they said, 'We have people in Kokomo who do that.' So why the hell aren't we working with Kokomo?" Doug and Ron agreed. They needed to see if the engineering department at Delphi's Kokomo facility really differed from the one at Flint. They had to be sure, because if they picked Delphi but had to pull out later, said Doug, it would threaten Dean's relationship with J. T. Battenberg.

"Frankly, FIRST isn't my priority," said Bill. "Ginger is."

"Dean integrates his priorities," said Doug, "and FIRST is part of it."

Ron envisioned maddening delays as the team's requests glugged through Delphi's labyrinthine system. But Doug believed that if Delphi could be pushed and managed, the company's vast resources would ultimately help make Ginger a great machine.

They enlisted Don Manvel to use his contacts at Kokomo to soften up the expected opposition. After all, why should this huge corporation loan a team of its best people to a small obscure group working on a secret project? Not an easy sale, and Don didn't make much headway at first. Then he visited Delphi with Dean, who mesmerized a roomful of top executives with one of his tent-revival performances. Suddenly Delphi's walls seemed about to come tumbling down.

A few days after that visit, hoping to clinch the deal, Don, Doug Field, Bill Arling, and Phil LeMay were on the speakerphone with some Kokomo executives led by a man named Bill Gray. Gray reported that Delphi's executive committee had agreed to loan Ginger some staff, and he explained the company's process for doing so: Delphi would hold a "sanctioning body event," during which a committee would consider "nineteen deliverables." The Ginger guys traded amused looks.

Doug asked how long this process took. Three weeks to select a team, said Gray, plus another four weeks of transition before they could start. The Ginger guys rolled their eyes at each other. If dot-coms operated on fast forward, the auto industry was slo-mo.

"That's two months," said Don, "and we don't have it." Gray hmmmed.

"We want the best of Delphi," said Don, "but we're offering the opportunity of a lifetime." Don had learned the Kamen formula: Make an outrageous request followed by a glowing promise.

"It'll be an amazing experience, I guarantee you," said Doug, jiggling the bait.

The speakerphone was silent.

"We even have an author writing a book about it," said Doug, still jiggling. Doug explained that over the next few weeks the dashboard team would be considering concepts and purging all but two or three, so if Delphi wanted to participate, they had to jump in fast.

"It sounds like the design competition we have when we get our FIRST kit," said Gray. His voice warmed: "The Kokomo High School TechnoKats. How often do you see Dean?"

"Dean is available twenty-four hours a day," said Doug, sensing a nibble.

"I've seen him at some FIRST things," said Gray, "but he probably wouldn't remember me."

"He might," said Don. "So Dean is around, and your people would get exposure to him," he added, twitching the bait again.

"This is exciting," said Gray. He sounded energized. He said that instead of doing this "logically," he would try it the DEKA way, which made the Ginger guys smile.

Don set the hook, asking for a meeting this week.

"You mean next week?" asked Gray.

"No," said Don. "This week. Tomorrow."

"Wow. Well." Gray hmmmed a little more. "Maybe I could do some things tonight, maybe show you some people."

"And we'll say yes or no," said Don, flouting the adage about beggars and choosers. "So when would you like to talk again? Later today? Tonight?"

"How about tomorrow morning?" said Gray, still catching his breath. "Eight o'clock our time. This is exciting. We have some brainstorming to do." He hung up.

Am I wrong, I asked Don, or did Delphi just give you everything you wanted? He smiled. "This is called supplier management," he said. "But it may not have happened if he hadn't had the FIRST experience. He knows what we're talking about."

Dean's wide contacts through FIRST often spilled into his business. "It's like Dean spreads this virus," I said, "and you never know where it's going to pop up."

Phil LeMay and Bill Arling morphed into a comedy team:

"I've got the *DKs!*"

"Yieee! The Kamen virus!"

"Outbreak!"

They felt lucky to have an ally at Delphi like Bill Gray. "But he may get in trouble," said Phil. "He's going outside the Sanctioning Body." Phil began speaking in the sonorous voice of doom: *"You've defiled the Seventh Deliverable of the Sanctioning Body!"*

"I kept imaging them all standing around in robes, muttering in Latin," said Bill, then switched into a singsong ecclesiastical voice: *"Body of Delphi . . . Aaa-men."*

Phil laughed so hard he had to bend over, his face red as an apple.

The search for a regulatory vice president wasn't going well. The first two candidates had been alarmingly bad. Dean, Mike Ferry, Tobe Cohen, and Matt Samson had just finished interviewing the third. All three marketers voted to reject her on several counts: She talked too much, said almost nothing, and was a lawyer. Dean considered the description redundant. "A lawyer's job," he quipped, "is to give you the least amount of information in the most amount of time."

He seemed willing to consider her out of sheer desperation, but the others objected strongly. "Her content per sentence-unit was pretty low," Dean admitted. "OK, forget her. Boom. She's dead. We need someone who can either get around all the regulators *or* figure out how to finesse them all, *and* who can answer a hundred specific questions about detailed regulations. So we need a strategic thinker and a grunt."

He moved on to Ginger's name. The leading contenders were now Flywheel, Relay, and Baton (the company itself was still called Acros), but Dean had had another brainstorm over the weekend. He narrated a long story about the thought process that had led him to the name, then paused dramatically, like any good storyteller, and delivered his kicker: Link.

He started selling the name to the marketers. Link was simple and worked as a noun, a verb, and a story. "When Ted Koppel says, 'Why do you call it a Link?' I can say, 'It's the *missing link* in transportation, between downtown and the suburbs, between the subway and work, between home and work.'" He looked around. "Well?"

"I like it," said Tobe.

Dean shivered in mock horror at pleasing a marketing guy.

"Does this mean we have to give the $70,000 back?" joked Tobe, referring to Lexicon's fee. Mike and Matt liked the name too.

Dean asked them to research link.com and the word *link* in foreign languages. "See if in Swahili it means 'rape your mother.'"

But the world turned out to be crowded with Links, many of them related somehow to vehicles. Another name bit the dust.

In late August the marketers invited a branding firm called Jager Di Paolo Kemp (JDK) for a meeting with Dean. Two of JDK's principals, Michael Jager and David Kemp, came to make the pitch about their branding philosophies: the Living Brand (defined on their baffling Web site as "the humanized brand character that creates lasting relationships with people on rational, emotional, and experiential levels") and their design philosophy, Consciousness of Chaos.

Everything about JDK was exquisitely, self-consciously cutting edge; Dean, by contrast, disregarded edges or jumped off them. JDK applied its Consciousness of Chaos to invent impressions; Dean applied physics to invent products that improved the world. JDK cared about image, Dean about substance. They were bound to clash.

After half an hour Dean had heard enough. He began with an assault on JDK's business card, an illegible avant-garde agglomeration of graphics and colors, made even less decipherable by being encased in a chic blue plastic sleeve. This cutting-edge *objet*, this impenetrable Living Brand that exemplified Consciousness of Chaos but not an escape from it, irritated the hell out of Dean. He did a little riff that bristled with the words "pretentious" and "useless."

Once again Mike Ferry was embarrassed—by Dean, not JDK. Mike wanted the agency. He would mend fences and explain how to approach Dean. Be factual, not abstract. Be conversational and avoid elaborate presentations. Be clear about how you would spend his dollars. Ignore his barbs. Dean often kicked like a wild stallion, but Mike was determined to get Ginger branded.

"We *have* to squeeze the funnel." Ron Reich was revving at Scott Waters. "We have to do things quicker, with less attention to fine detail. You're not going to have time to linger over details."

"Please don't make assumptions about my time," said Scott in a clipped voice. "We can't shortcut the process. Last time, you made the schedule, and whatever was left at the end was for industrial design. We can't do that this time. Understand?" His face had reddened as he spoke.

Ron raised his eyebrows. "Yeah," he said, throttling back a little. "But we have to do things quickly." Scott stalked off.

The Metro had Scott spooked. He ticked off the list of components he needed to redesign: the dashboard, the control-shaft base, the wheel and tire, the fender. "Basically everything," he said. At the moment, he was working on the control-shaft base. He wouldn't let himself think about the dashboard yet. Too scary. They hadn't even decided what information the dash would display, or how to display it.

For inspiration Scott had made a collage of images surrounding the words Metro and Pro. A hummingbird hovered near Metro, an eagle soared near Pro. The collage's Metro section had images of a subway, skyscrapers, city streets, a small apartment, commuters. The Pro side showed trees, trails, campuses, country roads, mountain bikes, a small town. Photos of sleek products occupied the middle ground—a coffeemaker, a Swiss Army knife, flashy sneakers, a high-tech bike, an iMac.

Scott had gotten his way, sort of, about adopting a better process for industrial design. For the new handlebar and other components, Doug had agreed to let Scott and Tao Chang generate a funnel full of ideas, then strain out all but two or three. They would do detailed studies of these while consulting with Ron Reich and Mike Martin about structural requirements and with the manufacturing group about assembly requirements. Finally they would pick a concept and go into full design mode. But Doug insisted that Scott and Tao complete the drill in double-time, which meant many late nights and tense conversations, like today's with Ron.

A few days later, at a meeting in the War Room about design decisions, Doug passed around an electric razor. Its small round display was a possible model for Ginger's dashboard. Dean dropped in and sat down. The topic turned to sidewalk friendliness. Dean wanted all collisions to be no worse than getting jostled by a pedestrian's elbow. He recommended soft bumpers again, the Nerf Ginger that wouldn't die. Doug sidestepped the subject.

They moved to the steering mechanism, which the engineers called the yaw. They had tried twenty-five or thirty methods, all kinds of crazy things. One of the nuttiest involved independent foot pads that resembled dual gas pedals—press the left pedal to go left, the right to go right. You had to jump on and off. "*Very* entertaining to ride," said Doug. Another prototype used "knee steering." The horizontal stem of a short T-bar hit the rider at the knees; the rider steered by leaning a knee to the right or left like a surfer. Only one engineer ever mastered it, J. D. Heinzmann, and he still proselytized for it every chance he got.

Two steering methods had survived: a switch on the handlebar that the rider pushed right or left with a thumb, and a twist grip that the rider turned left or right with a hand. Most of the engineers favored the thumb switch because it had been around longer, but for reasons of safety it was losing ground to the twist grip. When bumping over rough surfaces, a rider could easily lose contact with the thumb switch. It was also clumsy to use with gloves.

Both mechanisms had always been operated by the rider's right hand. The marketers had been pushing to offer left-hand steering as an option. The engineers hated the idea, partly because steering had always been on the right, but mostly because the left-hand option added complexity of design and assembly, and therefore time and cost. The manufacturing group disliked it for the same reasons. When a design issue escalated into extreme contentiousness, the engineers would cry, "*Jihad!*" The term often popped up when the topic was steering. Now, in the design meeting, Doug asked whether it should be left-hand/right-hand swappable.

"No," said Tim Adams.

"Yes," said Dean.

"Motorcycles don't have it," said Tim.

"You mean those *vehicles*?" said Dean.

"You get my point," said Tim. "Give me an example of when it's necessary."

For sports, said Dean, so people could have their right hand free. Tim pointed out that every time they added an option, they increased the workload and endangered the schedule. Dean suggested one of his typical compromises, meaning he couldn't lose—either make left-handed steering an option for extra money, or make it standard so the rider's right hand stayed free. Doug postponed the decision.

But Dean wasn't quite finished. Another idea for the steering mechanism had come to him last night. As he walked toward the whiteboard to draw it, Doug looked pre-Deaned, so Dean turned and said, "Don't worry, you don't have to build this."

"Can I think about it?" said Doug, smiling.

"At night."

Dean drew a strange device that carried a rider. "It's a frog," he said, turning to Doug. "Imagine that you don't stand on a platform, but on a swing with springs at the corners, so you're on an inverted pendulum with a potentiometer." He drew a curved pendulum under the rider. Such a design would have better turning stability and suspension. "It becomes more like a sport, like skiing," he said. "And maybe you wouldn't even need a handlebar."

"No," said Doug, studying the drawing. "You have to have a handlebar." He jumped up and began sketching next to Dean's drawing. They dueled with vectors and equations. Dean eventually admitted that a handlebar would be necessary, though not at low speeds. Doug conceded the point. Ignored, Tim and Don Manvel had drifted out of the room.

"The problem is that laterally we are imprisoned by static stability just like a car," said Doug. "Ginger can't do this." He hopped sideways.

"When you lean forward," said Dean, "Ginger says, 'I'm not going to let you fall. I'm going to run these wheels out in front of you.' It would be nice if Ginger could do that laterally as well."

Doug turned to the board and drew a few variations of Ginger. They stared at their sketches of what Ginger might become. "You try out a hundred of these to find one good one," muttered Dean.

Because of Dean's ban on consumer testing, Mike Ferry and his team of marketers had a maddening itch they couldn't scratch. They wanted to learn how and where and when people would use Ginger, what people would like and dislike about it, which features were crucial or negligible. But the biggest reason for testing was to do a "disaster check"—to find out whether people were willing to spend a considerable sum for such a machine. Since Dean had no doubts whatsoever about this, and since he was paranoid about secrecy, he ignored the marketers' wishes. He wanted to wait until at least January 2001 to do a small group test and to delay deep market testing until three to six months before launch.

This time line appalled the marketers and Tim Adams. They found unexpected allies on the board. Even Bob Tuttle was leaning in their direction. They met with Tuttle in late August to discuss it.

Tuttle's first item of business was a message from Aileen Lee, John Doerr's associate at Kleiner Perkins. She had forwarded a message that a friend had found on a listserv for Stanford's School of Business. The message had asked if anyone knew about a "stealth mode spinoff" called Acros that was developing a "new transportation technology" with a launch in May 2001. The poster had gleaned the information from a friend interviewing for a job at Acros. For Tim Adams and Mike Ferry, this confirmed how viral the Web could be.

The Stanford message was more of a droplet than a leak, but it worried Tuttle, probably because it worried Dean. Tim and Mike suspected that it traced back to their ad in the *Los Angeles Times* for a regulatory person. But the wall of secrecy looked increasingly porous. Tobe Cohen pointed out that the marketing team had been developing a Web site, which could expose the company, but the site was necessary to build the company, so it was a trade-off.

"It's dangerous," said Tuttle.

"The clues are already out there," said Tim. "The iBOT, DEKA, the patent for balancing." He didn't mention, though he had thought about it, all the times that Dean had told reporters that he had a secret project, and all the multimillionaire businesspeople Dean had invited to Manchester to ride Ginger, most of whom had disappeared back into their wide circle of contacts. "It's remarkable to me that fifteen months after *Dateline* we're still stealth," said Tim. Once again he recommended getting a public relations agency in place to handle the inevitable leak. But that would mean spending money for no clear return, which Dean was reluctant to do.

Mike Ferry and Matt Samson also complained to Tuttle about the obstacles to clearing people such as consultants, job candidates, and test riders. Even Tim, the CEO, couldn't authorize clearance. Dean reserved that power for himself, which meant that anyone who needed an authorization had to track down Dean and present his case, usually while walking to Dean's next destination at DEKA. This exasperating micromanagement wasted time.

Tuttle shrugged. Some things were unalterable. Moving on, he wanted to know how many Gingers the marketers thought the company would sell in the first year. The question was unreasonable. For starters, they hadn't been allowed to do any testing. Mike Ferry educated Tuttle about other crucial missing information: Volume forecasting was always difficult, but more so for a new product, and even more so for a "discontinuous" product that created its own category. Nevertheless, Tuttle wanted them to guess. Mike's crystal ball predicted 50,000 to 150,000 machines; Matt's 50,000 to 100,000; Tobe's 75,000.

Tim pointed out that they needed consumer information at least six months beforehand to plan the launch ramp. The testing itself would take five months. That meant it should start in January, six months outside of Dean's comfort zone.

They started brainstorming. Maybe they could do the early tests in secrecy, perhaps take over a movie set or a resort. Tobe thought they could "secure" a small college, but Matt shook his head—the media would sniff that out. Tim suggested a small European town interested in transportation solutions. Mike proposed a remote island in the Caribbean, though the likelihood that Dean would approve such extravagance was slimmer than none. Tuttle wondered if a place in Latin America would work, "someplace without freedom of the press." But no place seemed to meet the dual conditions of secrecy and significant population.

Later that day, Mike, Tobe, and Matt met with Dean. They showed him a video of the Ginger Advisory Board, a droll name for a few DEKA spouses, mostly women, who had recently been allowed to cruise around the lab on Gingers and answer the marketers' questions. First, John Morrell, the controls engineer, gave them the safety talk. Just after he warned that the machines were prototypes, so things might go wrong, the Ginger he was riding went crazy, spinning him around like a dervish. He jumped off and said, "See?" They thought he was kidding around. Otherwise the test went well. The woman who initially seemed most scared soon relaxed and said she felt safer than on a bike. Dean liked that. As for the two steering mechanisms, opinions among the riders split between the thumb switch and the twist grip.

Market research interested Dean, but he didn't consider it gospel. He liked to say that if you asked people where they would put a third eye, most would say on the back of their heads. But if you gave them the option of putting it on the end of a finger—he would wave his in illustration—the advantages were instantly clear. Dean had learned long ago that customers didn't always know best. They hadn't thought about the problem deeply enough to envision innovative solutions.

As August progressed, he had grown more convinced that a twist grip for the left hand was the way to go, despite people's natural inclination to want right-hand steering. "You can brilliantly document bad data," he told the marketers. "People are soon going to want their primary hand free. Kids who play baseball would probably prefer to catch with their right hand, but then they couldn't throw, so they learn to catch with their left."

A few more shots would be fired over steering among Ginger's engineers, but this jihad was over.

At 10:00 that night, Dean, Tim Adams, and I went to dinner at T-Bones. For some reason the conversation turned to Hiroshima and Nagasaki. Dean didn't understand the critics. If someone dropped a bomb on his house, he would kill the person if he could. "In war the point is to win, period," he said.

That reminded him of the time his grandmother gave him a bike when he was about twelve. Ecstatic, he rode it around the block again and again, until an older boy jumped out of the bushes and knocked him down.

"I was very small," said Dean. "He was a huge Aryan-looking kid, in high school." He told Dean not to ride his bike there anymore and added, "We don't like Jews around here."

Though terrified, Dean didn't stop thinking. "Finally I said, 'You'd better kill me now, because if you don't and you keep bothering me, my father has a pitchfork, and sometime when you least expect it, I'm going to jump out of the bushes at *you*, and I'm going to put that pitchfork through your neck.'" The bully backed off. To pound home his point, Dean rode his bike around the block another dozen times, until he couldn't pedal anymore.

"And I didn't get over it," he said. "I just got madder about it as the years went on."

It was August, and some of the team members were off on vacations. Dean never took one—he referred to it as "the V-word." For him relaxation meant machining a few parts for his flywheels or curling up with a technical manual or a physics text. The idea of lolling on a beach, or taking long walks, or indulging in days of undirected activity struck him as a deliberate dissipation of energy, volitional entropy, inconceivable waste.

The profile of him in *Wired* magazine had come out. As always, Dean was disappointed that the story contained more information about him than about FIRST. Cynics will find that disingenuous. Like anyone with a strong ego and a streak of the theatrical, Dean did love attention from the media and couldn't resist talking to reporters. But he also harangued them to write about FIRST and truly was puzzled, if flattered, when they focused on him instead. He wondered why anyone would be interested in such personal information, because it struck him as trivial. "If you read about physics," he noted, "you *learn* about physics." And what kind of person would rather read about people than about thermodynamics?

Jack Hennessy was irritated because the article referred to him only as "a banker from Credit Suisse First Boston" but mentioned John Doerr by name. Hennessy resented being described as a mere banker and blamed Dean, who felt terrible even though he had no control over the reporter's wording. Dean didn't tell me that another thing had irritated Hennessy even more: He didn't think Dean should have talked about Ginger at all.

As August ended and autumn approached, sidewalks were never far from Dean's mind. A few places had started banning scooters from them. Tim Adams was concerned. Dean professed not to be. That wasn't true, but his habit was to transform worry into unassailable confidence, especially when confidence didn't fit the facts. The laws of physics couldn't be defeated, but Dean considered everything else surmountable. In the past weeks I had listened to him hone the arguments he intended to make with regulators. He contended that the recent scooter bans were actually "a road map." If they ban them because they're too fast, he said, we'll go slower and say we're not like that. And if they *don't* ban them, great. Either way, we can't lose.

Ginger still didn't have a regulatory person, but a consultant had told the team's managers that after a preliminary survey, the machine didn't seem to fall under any state's definition of bicycle or "electrically assisted bicycle." But that didn't guarantee that Ginger could escape regulation. New York and Pennsylvania, for instance, had tough laws about bikes on sidewalks. "Motor bicycle" and "motorized bicycle" were grayer areas. In short, the laws, statutes, and definitions were a morass. The consultant had her own opinion about putting Gingers on sidewalks: "The idea of having these big things coming up behind me," she said, "scares me to death."

Mere opinion, definitely surmountable.

chapter twelve

The Path of Minimum Pain

September 2000

Before D1, Ginger's chassis had been jury-rigged from a dozen parts. Now it was cast as a single piece. The gyro mechanism had been condensed from ten parts to one module. Two microprocessors once controlled the motors, but now a single processor did the job, thanks to a new integrated circuit from Texas Instruments that was smaller, cheaper, and more powerful. J. D. Heinzmann expected other pieces of Ginger's innards to shrink and merge as new technologies became available.

In early September, J. D. was tinkering with the power loop, figuring out how to regulate those Siamese twins, voltage and temperature. When a rider demanded power, the temperature in the transistors spiked. Their silicon made them sensitive to heat. Ginger's transistors were rated at 175 degrees Celsius, but for safety J. D. wanted to keep the temperature well below that. If the transistors fried while a rider was zinging along, bad news.

So J. D. had been measuring temperatures in the transistors and assessing how fast they heated up. Pretty fast: 125 degrees Celsius in 4.5 seconds, close to his safety limit. Whenever the temperature climbed higher, he wanted to cool things down by automatically decreasing the voltage available. In other words, the rider would not be allowed to ask for more and more juice. The machine would have a speed limit. J. D. explained that he was essentially turning the transistors into valves that let current flow or pinched it off.

Was it hard to solve? "Yes!" he said, with a big smile. He was happiest among the thorns.

Limiting speed entailed a constant conversation between transistors, processors, and software. Once J. D. had worked out the voltage/temperature relationship, John Morrell and Jim Dattolo had to adjust the software to preserve something called "headroom." J. D. explained the term with a story. One day he was strolling with his young niece, who was holding the leash of an unruly dog. The dog sprinted off, pulling her until she couldn't run any faster. But she didn't let go of the leash, so her upper body began to tilt forward until it was too far in front of her legs, and she crashed. "It was horrible to watch," said J. D., "but a perfect illustration of running out of headroom"—that is, a reserve of power for acceleration.

J. D. and the software guys had to ensure that Ginger's motors always conserved enough headroom to get the wheels back under a leaning rider. This involved a paradox: Ginger had to speed up to slow down. When the machine received the command to decelerate or stop, its first response was an infinitesimal spurt of acceleration to move the wheels up under the rider's mass, so that he was no longer leaning forward. But a rider who pushed the motors to their max lost the ability to accelerate—he ran out of headroom. J. D. and John Morrell couldn't allow that to happen. A rider leaning forward aggressively would heat up the transistors and reach the speed limit, causing the computers to reduce the voltage and slow the machine down. But first the motors and software had to run the wheels back under the rider by briefly overriding the speed limit to access the reserved headroom and accelerate back into balance. In microseconds. It was thorny enough to keep J. D. happy for a while.

One morning in the first week of September, Dean walked around DEKA with an assistant who held a box of envelopes containing the long-awaited shares in Ginger. Everybody at DEKA, not just the Ginger group, was getting an envelope. Dean loved doing it and was feeling expansive. He ran into Benge Ambrogi, one of the early designers of Fred. "Some people are getting a bigger piece than others," Dean told Benge, "because they've been here longer or because they worked on Ginger or because they worked on Fred and helped to make Ginger possible. You fit *all* those categories." He handed Benge a sealed envelope. "It may just be wallpaper," said Dean, "but if the bankers are anywhere near correct in their estimates, you just became a very wealthy man." Benge grinned.

Over at Ginger, my informal poll found that most people were satisfied with their portion of shares. I asked Doug if he was content with his. "Content," he said flatly, measuring the word and doing a risk/benefit analysis on various degrees of frankness. "I have spoken to Dean about it," he finally said, "and I have accepted it. I don't really want to think about how my share compares to other people's. It's much more important that the organization has done this than what my share is."

"You don't think about the financial possibilities for yourself?"

"I'm focusing on the work to be done."

A good soldier. Maybe Doug didn't want to think about the size of his share, but it was clear that he *had* thought about it, and equally clear that he had expected more. Other than Dean, no one had been a stronger influence on Ginger, and no one, including Dean, could claim a stronger hand in its evolution. Maybe Dean resented that, or maybe he really believed, as he had once said, that Doug shouldn't expect a big share, because he had already been working at DEKA and should feel lucky to have fallen into his dream job. Nonsense, and more nonsense. But Ginger *was* Doug's dream job, so he accepted his envelope and soldiered on.

His role had changed in the past weeks from, in his words, "process managing to tactical managing." His list of more than a hundred "enterprise decisions" epitomized this change. Every item on it, he said, tapping the list, was a little fire he had to douse before it erupted into a blaze. He was doing more cross-organizational managing now as well, with Tim Adams, Don Manvel, and Mike Ferry. "I'm knocking down roadblocks rather than creating," he said.

Did he enjoy this phase? He didn't speak for several seconds. "I prefer the front end, when you're doing a lot of creating," he said. "But *not* to do this now would be more painful than doing it. So this is the path of minimum pain."

This phrase captured the project's current stage. Crazy frogs and cheap parts from Radio Shack were things of the past. Ginger was getting older, more expensive, more set in its ways. The team had to find the path of minimum pain and stay on it.

The atmosphere at Ginger felt less exuberant than during the D1 build. People seemed more serious and hunkered down, but a quiet resolve sustained everyone. Doug considered this a sad yet necessary development. Building a single prototype, he said, was far different from building fifty machines, as they were now doing for D1B, in preparation for building even more D2s. "It's gotta be like landing planes on an aircraft carrier," he said. "It has to be down pat, almost boring, because the stakes are so high."

Doug continued to fret about recruitment. To meet his deadlines he needed more engineers. In the hallway a few days earlier, he had tried to convince Dean to let him hire an experienced candidate who, in exchange for uprooting his family, wanted $75,000 a year and a look at the project he would be working on. Doug regarded both requests as reasonable, but Dean wanted to pay less and to show Ginger to the candidate only after he took the job and moved to Manchester. It was the same old battle, but the stakes had risen. Dean urged the engineers to speed toward production, then acted like a gremlin.

Meanwhile, progress on the dashboard design, which the engineers called the user interface, or UI, buoyed Doug. D1's swollen "football" dash was cluttered with five lights, two buttons, a key switch, and a power switch. Much of this had melted away in Scott Waters's latest renderings. Thanks to a suggestion from Phil LeMay, the key switch, power switch, fast/slow button, and security function had merged into an iButton—a small, round electronic key—which also functioned as a data port. The "follow mode" button, used to pull Ginger up stairs, had migrated into the steering mechanism. Simpler, cleaner, smaller—Doug's design mantras.

But there was a long way to go. I watched Doug and Scott work on some sketches of the dash at Scott's drafting table. All of them showed elliptical shapes the size of a watch face, perched at the joint where the control shaft met the handlebar. Bill Arling sauntered over and leaned on the table. Bill often spoke in condensed pellets of language. He would respond to an opinion with a sharp, "Reasoning?" Someone's reference to a report might elicit, "Contents?" I also liked his all-purpose, "Purpose?"

Now he considered the sketches and said, "Four by two." Doug and Scott glanced up at him. "The size of the UI board," said Bill. In inches.

They looked at him silently, looked back at their small impossible sketches, and slumped a little.

"Party pooper," I said.

"Reality inserter," said Bill.

While the engineers intently rejiggered the design, the manufacturing team sweated over the assembly. Designs that looked elegant in the shop often revealed an unsuspected dark side in the factory. Components that married like yin and yang in the lab disclosed irreconcilable differences on the line. Ginger's design engineers didn't think much about the time needed for assembly, but the manufacturing team obsessed about it. Awkward stages that added fifteen seconds here or forty-five seconds there would cost the company dearly when multiplied by millions of repetitions.

So as Ginger raced toward production, the manufacturing team elbowed its way into the contest, looking for ways to smooth the transition from engineering to assembly. One of their prime principles was poke-a-yoke, slang for error-proofing the assembly. If something *could* happen during the manufacturing process, it *would* happen, explained Chip MacDonald, Ginger's plant-manager-in-waiting. Chip had once worked at a plant that made juice cartons with pictures of oranges nestled in green leaves. The task was simple: Match the coded card for ink color with the code on the huge roll of paper, green with green, orange with orange. Simple. But a couple of times every month, the workers ran big rolls with orange leaves or green oranges. Faulty poke-a-yoke.

Perfect poke-a-yoke made mistakes physically impossible. Each of Ginger's cylindrical motors, for instance, featured two ears for screwing the motor down. The ears lined up with the screw holes only one way. That was good poke-a-yoke. But it was still possible to put the motors in backwards, so the engineers were adding a third ear to eliminate all possibilities except one. Chip was grateful to Ron Reich for some preliminary poke-a-yoking, like removing all the possible left-right mistakes.

In a corner of the downstairs lab, Chip and the other manufacturing guys had thrown together a homemade "plywood plant"—essentially a test track—made of sawhorses and planking. At the start, not including subassemblies, there were 209 assembly steps. Way too many, because each one represented a point where something could go wrong. What number of steps did Chip hope to reach?

"One," he said. "The ideal thing would be to put all the parts in a box, shake it, and pull out a Ginger." A guy needs a dream.

He worried most about the thirty-eight screws inside the chassis, including thirty to hold down the two main electronics boards. "We have to reduce that," he said. "We *hate* screws." Screws took too long to fasten. Screws inevitably fell into the machine, where they rattled around or, worse, caused electrical shorts. Chip had asked the engineers to reduce the number of screws. Sorry, they'd said, impossible before launch.

"We aren't accepting that," said Chip with a tight smile. "We're going to come up with something on our own. Because it scares the hell out of me." He kept imagining a guy with five minutes left on his shift, dropping a couple of screws into the chassis and shrugging it off.

He had set up thirteen stations in the plywood plant. A laminated page above each station explained what should happen there. The stations were provisional guesses about the best order of assembly. Chip would revise them based on hard experience.

The first D1B took ten hours to build. Many of the engineers cruised by like gawkers at the scene of an accident. When they offered advice or made jokes, Chip grinned to himself, comforted by the thought that all of them would get their turns. He insisted that everyone on the team build a Ginger, for two reasons: The process revealed assembly problems and forced the engineers to wrestle with their own designs. Chip let them install motors incorrectly, strip screws, drop screws into the chassis, and use the wrong screws (there were two similar sizes—bad poke-a-yoke). He watched them cross-thread the wheels and listened to them curse as they struggled to line up all the screw holes on the fender. He let them discover that the O-ring on the control shaft was binding under the steel washer. To get through the wiring stations, one engineer had to remove the chassis cover three times. One of the gaskets looked square but wasn't, and had to be studied before putting it on. "Once they've got it figured out," said Chip, "we tell them the kicker: Now put it on upside down; otherwise it'll fall out." Overlong guide pins forced an assembler to jiggle a transmission halfway on, partially tighten the nuts, then pound the tranny with a rubber mallet until it was fully seated. That's when Ron Reich, while assembling his Ginger, bellowed, "We're *going* to change *that!*"

At especially frustrating junctures Chip would say, "Are you horrified yet?" or "Isn't that terrible?" Occasionally one of the veterans would answer something like, "Not really. You should have seen the wiring on C1." But later, if Chip complained in a meeting that a design was causing problems, four or five engineers would usually chime in, yeah, they'd had a terrible time with that.

Chip's list of Manufacturing Issues got longer and longer. By mid-September there were seventeen just for the dashboard module. A representative entry: "Spring Pins are extremely difficult to insert. Must be pressed in with arbor press and frequently bend upon insertion. Also, they are too long and must be ground down after insertion." What a weight of frustration those sentences carried. The redesign of the dash needed to eliminate such defects.

Chip constantly rearranged the order of assembly. He aimed to shrink the time per station from two minutes to thirty seconds and to adjust the stations so that each one required about the same amount of time. A month after that first ten-hour D1B, Chip's assistant built one in forty minutes. At the real plant, Chip expected Ginger's trip down the line to take six minutes, most of it spent driving screws into the boards. A new machine would roll off the line every thirty seconds.

That is, if all the suppliers met their deadlines. "It only takes one part to hold you up," said Chip. It was hard to poke-a-yoke your suppliers.

Dean, K. C. Connors, Benge Ambrogi, and I climbed into Dean's black Hummer at 5:15 one September afternoon, with an iBOT crowded into the back. Dean was going on a fishing trip. The Massachusetts Institute of Technology was hosting a reception for a conference on humanoid robotics, and Dean intended to hook some engineers for Ginger and the Stirling.

The traffic in front of DEKA was backed up several blocks before the main artery. "Look at this!" said Dean, peeved. He drove across the street into the access lane fronting the mill buildings, then rumbled up it, ignoring the one-way arrows pointing at us. The lane ended at an area landscaped with wood chips. Dean rolled over it to the main street, then onto the highway toward Boston.

He enjoyed his Hummer. One night, returning from dinner, he had suddenly veered off his long curvy driveway and bulldozed through the underbrush. He chuckled at the way the owner's manual described the vehicle as "a wolf in wolf's clothing" and explained that it didn't need airbags because the other cars on the road served that purpose. With its growling diesel engine and sinister profile, the Hummer was an impressive machine, and it fascinated Dean's engineers. But even they couldn't resist occasional quips about the discrepancy between Dean's gas-guzzling road pig, not to mention his private jet and helicopters, and his lectures about fuel consumption and pollution.

Hypocritical? Sure, but that's true of everyone to one degree or another, and isn't the same as being a hypocrite. Hypocrites don't inspire people for long or earn their lasting loyalty, as Dean did. Dean's contradictions sprang from his complexity. His mind was not hobgoblinned by foolish consistencies. He would have nodded at Walt Whitman's lines, "Do I contradict myself? / Very well then I contradict myself." Dean got more exemptions for his inconsistencies than did ordinary people precisely because he wasn't ordinary. His brilliance, charisma, and accomplishments attracted others to him. He generated excitement, so people excused his flaws as the price of sharing in his dynamism. Dean sometimes coasted on that sense of entitlement, as exceptional people often do, whether artists, politicians, or movie stars.

Now, roaring toward Boston, he began telling Benge about his new idea for next year's FIRST game—selling him, really. Benge asked what the FIRST staff thought of it. Benge often volunteered to work on FIRST and knew that the people there always had their own ideas, which Dean usually rolled over and replaced with some devilish invention of his own. "Oh, they pissed all over the idea," said Dean, "which means that it's good."

That was typical. Criticism seemed to reinforce his supreme self-confidence, which sometimes swelled into arrogance. But as he liked to say, ideas that everyone liked were probably too ordinary.

On the other hand, extraordinary ideas could turn out to be south-pointing chariots. Tonight's reception took place at MIT's Artificial Intelligence Lab, a hive of futuristic engineering. Guests marveled at the COG, "an upper torso humanoid" that processed information visually and aurally with a human-looking hand, arm, neck, and eyes. Elsewhere in the lab, a big blue eyeball tracked movement. A couple of Sony's amusing robo-dogs, called Aibos, toddled around.

Dean rode through the crowd on the iBOT, attracting attention in this throng of engineers. Between answers to questions, he would loudly announce, like a fairgrounds barker, "We're looking for good engineers. We're desperate." He would zero in on someone and ask, "What's your background? Do you know any exceptional students?" People took photographs and followed him like the Pied Piper to a stairwell to watch the iBOT climb up and down.

Dean gave out a lot of cards and collected some, which he passed on to Benge. Benge would sort the catch and throw most of it back, a job he did often. "We also get a lot of letters from inventors saying, 'Look at this,'" said Benge. "But we've never worked on anything from someone else. We just work on Dean's ideas, and that keeps us plenty busy."

We wandered down to the Leg Lab to see M2, a two-legged robot designed to mimic human balance and movement. Its complicated mechanical thighs, ankles, and knees were formed from wires, sensors, actuators, and processors. M2 had been the cover story in *Wired* that month. According to the magazine, the robot contained $90,000 worth of technology. It could balance, on good days, and had taken a step or two.

"They're solving the wrong problem," said Dean as we rode back to New Hampshire in the dark. "If you wanted to fly, you could flap your arms *or* you could build a stiff wing and push air under it. Instead of building the airplane, they built the flapping arm. It's true an airplane isn't as elegant as a bird, but a bird can't go Mach 3.1. There's a built-in limitation in the bird."

That was the flaw in everything he had seen tonight—it was ingenious but theoretical, academic. He considered the iBOT the world's most sophisticated robot, in a different league from the experiments on display at MIT. The technology there was called futuristic because, unlike the iBOT and Ginger, it didn't work *now*. It illustrated the difference between science and real engineering. On its best day, M2 was a south-pointing chariot. The iBOT and Ginger were compasses.

I drove Benge home from DEKA. On the way, I mentioned how happy he had looked to get Dean's envelope of shares. Suddenly agitated, Benge said that, frankly, he had expected more. He pointed out that because of the iBOT, which he had helped to create, Dean had collected millions of dollars from Johnson & Johnson. Dean had the jet and the Hummer and the helicopters and the mansion, whereas Benge drove an eleven-year-old Volvo and lived in "a lunch-meat house." "The contrast," he said, "is not lost on me." He acknowledged that the iBOT had been Dean's idea and that DEKA was Dean's company. He also believed, to an extent, that Dean's interests were the same as his. But he didn't feel that the envelope reflected his major contribution to the iBOT and to the technology that had migrated into Ginger. He said his complaint was more about recognition than money. Doug Field must have felt the same way.

Benge paused to collect himself, then added that the work and atmosphere at DEKA were so invigorating that he couldn't imagine leaving. He had applied to be an astronaut once, because even DEKA couldn't compare with shooting through space. But when he got rejected, he wasn't too upset, because he still had DEKA—Spaceship Kamen.

At D1's coming-out party the previous May, while everyone else drank beer and celebrated, one engineer had stood in front of the machine for several minutes, pensive. I asked what he was thinking about. "What we need to make better," he said softly. He also said, "I'm really looking forward to pounding one and breaking it as fast as possible."

His name was Mike Martin, overseer of Ginger's "mechanical integrity." His job had two parts: first, to warn the engineers and industrial designers about possible structural flaws in their designs, and second, to reveal those flaws through lab trials. He did the first job by testing a design against analytical models on his computer or consulting what Scott Waters called "Mike's scary books." This process became known as Martinizing. One of the first questions the engineers asked about a new part or design was, "Has it been Martinized?"

He did his second job by abusing real machines with diabolical tests until they broke. He was thorough and ruthless. He dropped, shook, bent, twisted, and thrashed parts. He wore out components by running them nonstop for weeks. If a test failed to defeat a part, instead of pronouncing the part OK, he devised a harsher test. "The two questions structural engineers ask," he said, "are 'Is it strong?' and 'How strong is it?'" He couldn't answer the second one until a part had broken.

Dean had invited Mike to DEKA four years ago from Enstrom Helicopter, where he had been chief structural engineer on the company's innovative turbine engine. At DEKA he had worked on the heart stent and then on the iBOT before shifting to Ginger.

He and Ron Reich sat next to each other, partly because Ron needed so many things Martinized. Mike was as quiet and mild-mannered as Ron was noisy and boisterous. Like J. D. Heinzmann and John Morrell, Mike was a natural teacher with a gift for explaining his work, which he did in clipped sentences. Aside from his shelves of scary books, his office décor included a small table made from a slab of Plexiglas atop a stack of tires.

Once the D1Bs began coming out of the plywood plant, Mike lived downstairs in his testing lab, off the main lab. To the engineers and industrial designers, this room evoked a torture chamber or house of horrors. Yet Mike's lab was essential for exposing Ginger's structural betrayals. The engineers knew that if their designs had flaws, Mike would find them. This was disquieting, as the possibility of exposed mistakes always is. But it was also comforting, because they knew that Mike had their backs.

"If the machine breaks a lot," he said, "we're dead. So anything that will kill the business has to be designed out." His catastrophic imagination ranged wide, from a vibrating capacitor that could snap a wire to the consequences of dropping Ginger 3 feet from a pickup truck. Onto one wheel. Better test that. In fact, he intended to drop-test Ginger from different heights and angles to make sure seals held and fluids didn't leak. He intended to run it through sand, slush, and dirt. Repeatedly.

"If it lasts forever, that's good," he said. "If it lasts ten years, that's acceptable. If it's less than that, we're worried."

One day he clamped a control shaft from Ginger into a device equipped with gears and pulleys. The mechanism looked like a medieval rack designed by a high-tech inquisitor. It would bend the control shaft five times per second. Mike expected to need one thousand to ten thousand cycles "to get a data point." That is, to break it.

About a dozen members of the team crowded around the rack to watch grim public entertainment. The control shaft was Bill Arling's part. He stood to one side, shuddering and grimacing. As Mike had expected, the glue joint near the top failed in a few minutes. On another test, one of the flared flanges on the control-shaft base cracked off with a *pop!* exactly as Mike's analysis had predicted. During lateral testing, the sharp lip of the inner shaft cut right through the outer shaft after about ten thousand cycles. That flaw, soon dubbed the Veg-o-matic, surprised Mike. The engineers, humbled, got busy redesigning.

Unlike his counterparts in, say, the auto industry, Mike couldn't draw upon decades of data to baseline his analyses. Ginger was the only machine of its kind. There was no data. No one knew how people would use it. Would they take it off foot-high curbs? How often? Would they run it at full speed for an entire battery cycle? On a beach? Would they bang it down the stairs or roll it carefully?

"If you have an infinite amount of time, you can get an infinite amount of knowledge," said Mike, turning an engineering fantasy into an adage. But shortages of time and data forced him to use fudge factors—educated guesses—in some of his calculations about strength. He guessed conservatively because he had to be sure Ginger was durable.

But he also worried about making Ginger *too* strong. Overdesign meant wasted expense. He decreed a spec of 300 psi (pounds per square inch) for the handlebar, but discovered that even the most aggressive riders never pushed the bar past 70 psi. Overdesign. He ran the tires on a drum machine for days at 25 miles per hour with a 250-pound payload. The tires did splendidly, which was great, but Ginger would never go 25 miles per hour. So should the designers remove a half-pound of rubber per tire—a tremendous savings when multiplied by millions of tires—and retest?

Overdesign also would make Ginger heavier than necessary, affecting range and performance. A lighter Ginger could go farther per charge. Or the team could opt to maintain the same range but cut costs by pulling out some of the batteries. Overdesign would eliminate those choices.

So Mike kept torturing Gingers in his lab, searching for the balance between strength and cost.

The team's managers couldn't agree on whether to build a battery charger into the Metro or make it a separate part. "A good decision is always preferable," grumbled Ron Reich, "but in my mind a bad decision is better than no decision, because at least it moves you forward so you're not just floating. Somebody needs to make a decision and let me off the leash."

Later that day at the coffee machine I ran into Scott Waters. For once, he agreed with Ron. The marketing guys wanted a built-in charger for Metro, said Scott, but there was no room in the only possible location, the control-shaft base. Nevertheless, marketing insisted. So Scott designed an enlarged control-shaft base, but it looked bloated, which neither he nor marketing liked. So Scott took the charger out. But marketing said, Well, no, we really do want a charger. Scott snorted. "We just need someone with balls to decide something," he said.

We headed back toward Ginger with our coffee. Scott had been working even longer hours than the rest of the team and his nerves were jagged. So many things needed his attention—the control-shaft base, the fender and wheel for Metro, the dashboard, the handlebar. All were major pieces of the machine. He felt bogged down in details when he desperately needed to be seeing these elements with fresh design eyes.

As we reached the middle of the glass skywalk that connected the buildings and collected the last of the summer sunlight, Scott stopped and the words poured from him. For some reason, maybe the warm light or the suspension high above the old Millyard, the skywalk seemed to open him up. "I've been having trouble sleeping at night," he said. "A *lot* of trouble. I wake up with night frights, thinking about everything. Everything always seems worse then, because you're half-asleep. I keep thinking I'm not good enough to do this."

He looked stricken. The river glinted below him. "I just told Doug that I'm not sleeping, that I'm having deep doubts about myself. He said in most companies you're given severe limits, but here you come in every day and there are no limits, so you really have the opportunity to see what you're made of, what you have inside, and so you're constantly hitting yourself against that, and after a while it takes a toll on you psychologically."

He looked young and vulnerable and passionate. He had put all of that intensity into the machine, along with his sense of style and fun, but sometimes his virtues blocked his own view.

"I just don't want to blow it," he continued. "I feel like this is my big chance to do something special, and I want everything on this machine to be right. When you're making a product in these volumes, I think you have a responsibility not to contribute to the clutter out there. There's *so* much bad product design, and I have strong feelings about not adding to it. I don't want to look up at the factory and see thousands of these and feel like *any piece* of it is bad. So I keep having these thoughts that another designer could do a much better job and wouldn't be having all these problems. I want it to be perfect. Doug and Ron tell me that no new product ever launches without things that need fixing. But I don't want to accept that."

He was still young enough to beat himself against the idea of perfection. He was growing up on Ginger, and the growing pains were severe. He admired Apple's commitment to industrial design, so I told him about the first Macintosh computer, which had been amazing, unprecedented—and deeply flawed. "Interesting," he said, nodding thoughtfully. "But I still have to come up with something great."

I was driving Dean to the airport. He was going to Israel with a group of businesspeople to meet with leaders there. Dean mentioned that John Doerr and Michael Schmertzler had made their joint proposal about funding the Stirling engine project. As expected, it had been ridiculous. They wanted to discuss it, but Dean told them that the gap looked unbridgeable, and he didn't want to risk their good relationship on Ginger by trying. Schmertzler, in a move reminiscent of Doerr, pinned the blame on the venture capitalist and suggested that Dean deal with Credit Suisse First Boston alone. But Dean was checking other options. He told Schmertzler, "If a bank seems like the way to go, *we'll* call *you.*" He also told Schmertzler that one of his co-travelers on the long trip to Israel would be the chairman of Chase Manhattan Bank. He knew how to play these guys.

As his mechanic pulled the jet out of its hangar, Dean said that he had invited twenty-six people from DEKA to a secret meeting at his house last night, using blind CCs so the recipients didn't know who else was coming. The meeting's purpose was "to fix the funk at DEKA—to get people feeling and thinking about DEKA and projects the way they used to." He looked troubled, as he always did when discussing his melancholy child. "We'll fix it," he muttered. "*We'll fix it.*"

Rumors about the meeting had leaked at DEKA, of course, and since Dean had tried to keep it secret, people's uneasiness festered. Evidently the stock envelopes hadn't been the restorative he had hoped. Dean couldn't put his finger on the reason for the malaise. Part of it was demoralization on the iBOT project, caused by endless manufacturing problems and government regulations. Another part may have been DEKA's continued growth and prosperity. In his book *The Tipping Point*, Malcolm Gladwell writes about Robin Dunbar's "Rule of 150." Dunbar's anthropological studies of villages, armies, and other social units suggest that when a group exceeds one hundred fifty, it begins to splinter. Some of its members remain strangers, clans form, and the group loses its cohesiveness. About two hundred people now worked at DEKA. Dean no longer knew all of them, and many of them didn't know each other. Dean had always been DEKA's dynamo and glue, with his nose in every aspect of every project. But these days he spent more and more time making phone calls and schmoozing investors, and less and less time in the labs. He couldn't keep up with all of DEKA's people and projects.

But he returned happy from Israel, where he picked up pledges for another bunch of FIRST teams. And Bill Sahlman had just sent him another possible investor for the Stirling engine project, Rebecca Mark. She had been Kenneth Lay's protégé at Enron before running one of Enron's now-notorious subsidiaries, Azurix, which had quickly lost millions of dollars. Mark had resigned as Azurix's chairman and CEO after selling her Enron stocks, reportedly for nearly $80 million.

She visited DEKA a month later, in late September, looking for something to do, and immediately saw the Stirling's possibilities. She offered to arrange investment funding within ninety days using her contacts in energy and real estate. The Stirling excited her, she told Dean, because in five years the large power utilities would have gone the way of Ma Bell. The power revolution would make the telecommunications revolution look like small potatoes. She urged Dean to combine the Stirling with Teletrol, his climate-controls company, for computerized power in residential developments.

When I walked into Dean's office on the day after Mark's visit, he had the phone at his ear, looking elated. He gave me a vigorous thumbs-up and switched to speakerphone. Rebecca Mark was saying that Morgan Stanley was "wildly excited" about the Stirling and thought that $30 to $50 million would be easy to arrange. She had told the bankers that she didn't yet know what her relationship to Dean was, she just wanted to do something fun. Dean ignored this hint. As they were signing off he said, "And look at that FIRST stuff." He hoped Mark would be his wedge into the energy industry for FIRST sponsors. "Resistance is futile," he said.

At dinner that night Dean ranted a bit about how the spendthrift Ginger team had added $5 million to the launch budget in the last couple of weeks. At this very moment five people from Ginger—five!—including Tim Adams and all three marketers, were in Tokyo splurging. Why did marketers have to travel in packs? Why were they staying at the pricey Imperial Hotel? The way the Ginger people spent money shocked and offended him, as if they were luxuriating in corporate expense accounts.

His mood improved when he started talking about a new project: next-generation Ginger. Tim, Doug, and the investors had been pushing him to get started on the idea. Tim had been urging Dean to put a new product in the pipeline every six months, but Dean had dragged his feet. He eventually put Benge Ambrogi and Larry Gray, another DEKA engineer, on the project. They had moved into the empty Ginger Penthouse, long since abandoned as too small. Dean also had asked Chris Langenfeld, the engineering manager of the Stirling project, to steal an hour a week to work on the future Ginger.

Larry Gray had designed the stent. Even though he hadn't worked on the iBOT or Ginger, Dean described him as such an incredible engineer that he would have "something up and running in two weeks." Dean considered Larry and Chris Langenfeld the two best pure engineers at DEKA, though others had different strengths. If he were on a deserted island "with nothing but a blowtorch," his engineer of choice would be Kurt Heinzmann. If there were a few basic tools, he would want Chris Langenfeld. If the island happened to have a machine shop so it was possible to build, prototype, and test, he'd want Larry Gray. If there were ten people who needed to be organized into a team, Doug Field.

He was confident that Larry, Benge, and Chris would devise a surprising new Ginger. "Today there's so much technology, the solution is right in front of you every day. Innovation," he said, "is a frame of mind."

chapter thirteen

All Hat, No Rabbits

October–November 2000

Doug constantly took the engineering team's attitudinal temperature, checking for problems. In mid-October 2000, a year and three months before the new launch date of late January 2002, he gathered his group in the War Room and wrote a series of open-ended statements on the whiteboard. He asked each person to complete the statements on stickies and then post them on the board. The responses offer a snapshot of the team's attitudes at this point in the project.

> *If I could change one thing in this place/program/product it would be . . .*
>
> —to add more software resources
> —more quiet time to work (two stickies)
> —quieter transmissions
> —outside riding time
> —we all work on one floor (four stickies)

If there's one thing I want to be sure doesn't change, it's . . .

—the team's camaraderie (ten stickies)
—the creativity that comes from interdisciplinary interaction
—the donuts (two stickies)
—the passion for this product (sticky in Doug's writing)
—Doug (three stickies)

What I miss about the way we worked six months/one year ago is . . .

—time to ride (four stickies)
—quiet time
—things were much tighter, we all pitched in to get it done
—beer Fridays—combined social/technical interactions
—frog-kissing

As I was reading these responses a day later, Ron Reich walked into the War Room, sucking a Tootsie Roll Pop. Doug had started keeping a jar of them near his desk, and the team often worked with white sticks poking from their mouths. "We had another soul-searching session," said Ron. "Doug had some concerns about morale. Frankly I don't think that shows up here. I *was* disappointed in one thing." He pointed to the category "one thing I want to be sure doesn't change."

"There were only three 'Dougs,'" said Ron. "He's the glue that holds this team together, no doubt about it. I know *I* couldn't do it without him."

I had heard the same thing from almost everyone on the team, and suspected that more people hadn't written down "Doug" because they couldn't imagine Ginger without him. He was the team's House of Gyros. Later I asked Doug what he had put under "What I miss most."

"Contact with the product," he said quickly. (Translation: riding Ginger.) "But that's the way things have to go, and that's my job."

In October Dean finally allowed Mike Ferry and his marketing team to take a small step away from stealth mode. Thirty-six outsiders, all vouched for by someone in the project, were going to ride and evaluate

Ginger. The marketers called the group the Boston Riders Panel. Two rounds of testing would be held outdoors at Bill Sahlman's secluded home near Boston. Another round would take place indoors at a Boston company funded by Michael Schmertzler's group.

Not many members of the Ginger team had ridden Ginger outdoors, so there were lots of volunteers to help set up the test at Sahlman's. Enthusiasm and immature software combined to cause several minor accidents. Chip MacDonald somersaulted over the handlebar but looked happy anyway because he had finally gotten to ride outside. Brian Toohey, who had just started as the person charged with proving to regulators that Ginger was safe on sidewalks, hit the pavement hard, tearing his pants, cutting his hands, and breaking his wrist.

The day after the accidents, controls engineer John Morrell was in good spirits. He felt vindicated. People sometimes treated him like a worrywart because he was always issuing warnings about faulty code and the machine's volatility. "I knew people wouldn't listen to me until somebody got hurt," he said.

The accidents proved to everyone what John already knew: The code wasn't ready. But whenever he typed in new math that changed or added something, everything else had to be recalculated, and he could never be sure what the results would be. If Dean or Tim or Doug or Mike Ferry told him to make sure the machine never caused riders to fall off, he could either drastically reduce the speed limit *or* write and test code for three months until it was perfect. But everybody wanted flawless new code *now*. He sometimes referred people to the sign on his office partition:

```
1. Fast
2. Cheap
3. Good

Pick 2
```

John had urged Doug to have a trustworthy EMT or doctor on hand during the consumer tests, but Ginger behaved and nothing went wrong. All the riders learned quickly and loved the machine. Mike Ferry would finally have some news to report at the upcoming

November board meeting. Technically, the tests showed the engineers that they needed an accurate battery-state estimator—that is, a way of telling how much current they could pull off the batteries at any moment. It wasn't too hard to solve. Most problems were getting smaller, right on schedule.

Benge Ambrogi, Larry Gray, and Doug Field were in the Penthouse discussing the advanced Ginger project. Doug suspected that Dean planned to yank the machine's future from his team. "There are a lot of guys over there," said Doug, referring to his engineers, "who don't want to just be supplying a plant in a year. They want to be working on new products." Doug didn't intend to give up Ginger without a fight. When the prototypes designed by Benge and Larry reached a certain stage, Doug wanted them handed off to his team for refinement. That was sensible as well as fair, since his team was better at product development than the DEKA guys. But he worried that Dean had other ideas. "The whole thing is a minefield right now," he said.

Dean told me he was withholding the new Ginger from Doug's group for two reasons. Doug's team, he said, was already "overworked and underloved," and giving them the advanced Ginger would increase resentment within DEKA. "And they've already got a little GM organization over there," he added. "They're not going to do anything fast and cheap. They have too much process and not enough product." Fast and cheap, the DEKA ideal. Yet $900,000 of the investors' money had been allocated for the advanced Ginger team.

Larry and Benge did make quick progress. By the first week of November, each of them had built a prototype. Both contraptions excited Dean. Either could become what he called "a poor man's Ginger." One autumn day while we were looking at Larry's contrivance, Kurt Heinzmann, the engineer who had helped build the first Ginger, walked in. Dean asked what he thought. Kurt gave it the once-over.

"I think it'll be hard to ride," he said.

"I think it'll be a gas," said Larry.

"It'll be like the bad old days," said Kurt, who longed for them. "We'll build it and then we'll say, 'Oh!'"—he hit his forehead—" 'How did we forget *that*?'" The wobbly process of invention.

Scott Waters was scraping at a piece of hard foam for the new handlebar design. Rock music boomed from his radio. He had sketched thirty handlebar concepts. The design team had voted and narrowed the possibilities to two. Both differed radically from the W-shaped bar on D1.

Scott picked up one of the foam prototypes and walked over to a D1 for a comparison. His new design flipped the sharp-angled W-bar into a lazy M, with handles pointing back toward the rider rather than horning toward pedestrians. Scott also had shortened the handles and angled them inward another 10 degrees, in a gentle curve. D1's steel handlebar would be replaced by a plastic clamshell. The new one looked sleeker, tighter, cooler. D1's dashboard, the swollen boll-on-a-branch that the engineers called the "football," had disappeared. It now seemed odd that anyone had ever liked the old handlebar.

I asked Scott if he felt more in control of things these days. "No, we're out of control," he said. "But maybe you saw that quote Doug had up from Mario Andretti: 'If you're not uncomfortable, you're going too slow.' "

A week later Scott was talking to Doug Field, Ron Reich, and Kevin Webber, a design engineer, about the shape of the battery charger. After one of Doug's suggestions, Scott, looking sly, said, "It'll be like that line curve on the Cougar."

Doug winced. "*No,*" he said. There was this line on the Mercury Cougar, he explained, "and when it gets to the quarter panel it just . . . *stops.* It's *awful.*"

Now that Scott had him going, someone mentioned the Pontiac Aztec. This time Doug looked more disgusted than offended. "The Cougar looks like it got hit with the ugly stick," he said, "but the Aztec looks like it got hit with the whole damn tree."

They were still trying to tweak an angle in the chassis to make room for the batteries. Kevin and Scott wanted another day on it. Ron looked annoyed and started growling at them. Doug asked if they could make up the time anywhere.

"We *can't* cut any more content," said Scott, glaring at Ron.

"Oh-ho, yes we can," said Doug, smiling.

Scott, unused to this from Doug, tensed up. "We *can't* design any faster," he said.

"Agreed," said Doug. "But that's not what that means. It means *stopping* designing and giving something to Kevin." Time to destroy, not create.

Doug felt confident about D2, which was on schedule. D3, on the other hand, the final "production intent" model, was slipping a little every day. He had warned Dean about it and didn't feel blameworthy. "I don't think there's any other group in the world who could have done this in the past year and a half," he said. "I sleep well. I can look in the mirror."

He still felt "massively constrained" by his shortage of embedded-software engineers. He was supposed to have four by September 1, but now it was late October and Jim Dattolo was still working alone. Dean and the investors didn't want to hear about slipping the schedule again, of course. Doug hadn't been able to hire anyone for two reasons. First, the competition for software engineers had gotten ridiculous. Desperate companies were paying candidates $5,000 signing bonuses. Current employees were getting large bonuses if they promised to stay for another year. But the other cause was Dean. He wouldn't approve high salaries, much less bonuses. Doug wasn't even allowed to tell the few candidates who expressed interest what they would be working on. In his recruitment ads, Doug used the lures he had:

> *We are developing a breakthrough technology that will provide an environmentally responsible solution to one of the world's most pressing problems!*
>
> **Do you want to change the world?**
>
> *Acros LLC is looking for talented, passionate, and unique individuals to design and develop this revolutionary electromechanical product that will create a new industry. . . . Our environment is where smart, technically competent people can thrive, expand their knowledge, and have the opportunity to develop products which will change the lives of millions. This is the project you will tell your grandchildren about.*

But software engineers weren't picturing their grandchildren, they were saying, "Show me the money." It was tempting to contract out the software work, but that would be foolish.

"When this machine comes out," said Doug, "Honda or GM can buy one and tear it down and figure out everything—*except* the software. That's our Coke formula. Strategically, it would be a huge risk to let it out."

Other parts of the company were growing fast. In June Acros had twenty-two employees. In September there were thirty-two. By the end of the year there would be more than fifty. New faces appeared every week, hired to handle suppliers, manufacturing, commercial sales, and e-commerce. More salaries meant a higher burn rate, which meant more pressure to meet the launch deadline. But unless Doug got more help to make the product, these new nonengineering employees weren't going to have much to do. Though morale remained high, Doug's team felt stretched thin. He had been getting e-mails posted from people in the lab at 1:00 A.M.

Doug wanted his engineers to feel slightly overworked, a lesson he had learned from Tim, because a smart person with time on his hands could instigate extra work for fifteen other people. Better to give his team too much to do and force them to get creative and make choices. "But we're way beyond that now," he said. "We're at the point of choosing which child will eat, D2 or D3."

He didn't think Dean really believed that D3 was in danger, because scares like this happened all the time at DEKA. "The way DEKA works is all-nighters and pulling rabbits out of a hat," said Doug. "Dean is always an optimist. But this can't be solved by all-nighters. There aren't any rabbits in the hat."

One night in October, I went to dinner with Jeff Finkelstein, the president of a small Vermont firm of consulting engineers called Microprocessor Designs. Finkelstein had known Dean for a dozen years and had worked on several projects, including the dialysis machine, Ginger, and the Stirling engine. He stayed at Dean's house while working at DEKA and was one of the few people outside the company whom Dean had given an options envelope. Like Dean, Finkelstein was blunt and opinionated.

At dinner, sounding just like Dean, Finkelstein complained that the Ginger team burned too much money and wasted too much time on charts and meetings "and a lot of happy horseshit." With the exception of J. D. Heinzmann, Finkelstein considered the team lethargic nine-to-fivers. "And that is *not* engineering." I didn't recognize his description of Doug's group.

Finkelstein had noticed another thing that irritated Dean. "He hates it that Ginger wants to be separate and not share employees with DEKA. That really twists his nose." Finkelstein thought Dean was foolish to bring in Delphi Automotive Systems, a big slow bureaucracy, and he called Dean's mania for secrecy crazy. He had watched Dean change over the years. The two of them used to go for bike rides after work sometimes, or out for a few beers. "No more," said Finkelstein. "Now all he thinks about is work. He doesn't really have . . . *fun* anymore, the way normal people do. He used to ski and do things. Now everything is work."

Finkelstein had noticed changes at DEKA, too. It was once a smaller, tighter group that worked longer hours with higher energy, and Dean was involved with everyone. "Now he holds himself apart a little more." Maybe that and the Rule of 150 helped explain the DEKA funk.

For the moment, Dean wasn't concentrating on the funk, but on the mid-November board meeting for Ginger. Michael Schmertzler wanted to snuffle in the details. He liked to know if Acros was getting 4.05 percent or 4.10 percent on the overnight sweep account, said Dean, whereas John Doerr was at 50,000 feet and looking up. Doerr had told him that his only issue for the board meeting was getting Ginger onto sidewalks. If they could do that, Doerr said, they would be the highest-cap company on the NASDAQ in three years. Dean considered both men very smart in different ways, but he identified more with Doerr.

For several weeks, Dean had been refining his oral arguments about getting Ginger on sidewalks. If local regulators banned scooters on sidewalks and limited them to bike lanes, Dean would say, Of *course* scooters belong in special lanes, they're *vehicles*. Then he would ask the regulators about their rules for pedestrians. Were people allowed to run on sidewalks? Were they allowed to crawl? Skip? Wear a signboard? Could they wear spike heels? After all, these could be as dangerous to pedestrians as Ginger.

Today he was directing his monologue at Brian Toohey, the new vice president for regulatory issues. When Brian had flown to DEKA for his interview, he had no intention of leaving Washington, D.C., for Manchester. The words "stealth project" had aroused his curiosity, that's all. But then he met Dean, who sunk the harpoon. Brian was supposed to catch an afternoon plane but tarried for dinner, then stayed up talking at Dean's until 2:00. By the time he went to bed, he knew Manchester would be his new home. A month into the job, he was still getting used to all the changes.

The task ahead of him looked formidable: convincing regulators and legislators that Ginger could safely coexist on sidewalks with pedestrians. The timing was against him. A surge of injuries caused by skateboards, push scooters, and motorized scooters had led some states and communities to restrict their use. But Brian didn't seem worried. A big friendly man in his early thirties, with a booming voice, he already knew how to massage the bureaucracy. He had started his career with the U.S. Department of Commerce, working for six years in Eastern Europe. Next he spent four years at Iridium, Motorola's ill-fated attempt to create a global mobile communications satellite consortium. He had been responsible for worldwide licensing and had succeeded with regulators in almost every country. He came to Ginger from Air-Cell, which provided air-ground communications for general aviation and required a lot of lobbying at the Federal Communications Commission. Incidentally, he spoke five languages. He had left a serious girlfriend in Washington, and as a joke I asked if she had started at DEKA yet. As a matter of fact, she had just visited Manchester to talk to Dean (he hired her). Dean was ecstatic about Brian, even after he broke his wrist riding Ginger.

"I'd like to develop these arguments up the ying-yang," Dean was telling him, "so that when I sit across from regulators who say, 'Of course it's a vehicle, it's got a motor and wheels,' I can say, 'Oh, so if my leg was blown off in Vietnam and I have an artificial leg and it has a motor in it, am I a vehicle? If I put a wheel on the bottom of my artificial leg, am I a vehicle?' We want them to think, 'If I don't get out of this conversation soon, I'm going to look like a Luddite.'"

He was just warming up. If the regulators gave up on definitions and moved to arguments about safety, he would say, If I step on your toes, I can break them, but Ginger rolls over feet with no problem. If I kick you, it hurts, but if Ginger hits you, it doesn't, because it's soft.

Right, chimed in Brian, we'll develop arguments to get exemptions.

No, said Dean, that's setting the bar too low. We have to be aggressive and take the position that *we have the solution,* and if you don't want it, we'll go elsewhere. We want them to *help* us make their cities Ginger-friendly by getting rid of traffic. We expect merchants to *pay* for Ginger kiosks. "We are going to make the pedestrian king again," said Dean. "The threshold isn't will they let us do it, it's what will they do *for* us."

In late October, Dean and Tim were prepping for the next day's meeting with Credit Suisse's Michael Schmertzler and looking ahead to the board meeting two weeks later. Tim was pushing Dean to let regulatory officials on board early, by the end of this year or the first quarter of 2001, despite the risk to secrecy. He also wanted to bring in someone who understood the Consumer Product Safety Commission, so that agency didn't become an obstacle later on.

Dean put it all into his mental processor, but he wasn't in a rush to show Ginger to any bureaucrats. He gave Tim some news about Delphi's new lithium polymer battery. The company wanted the battery contract for the iBOT, but had asked DEKA to pay millions for the tooling. Negotiations had dropped that to one-quarter of the original price. With that figure on the table, CEO J. T. Battenberg and Donald Runkle, executive vice president and second-in-charge, had visited DEKA to make a play for the battery contract. Dean, operating on instinct, decided to risk showing them Ginger. They had the usual dumbfounded response, leading to another late night full of talk and plans. And now Dean had gotten an e-mail that there would be no charge for the battery tooling.

"I've never heard of something like that," said Tim.

"In other words," said Dean, "J. T. said, 'I *want* this business.'"

Next Dean wanted to push Battenberg about gyros. Right now Delphi's were too big, noisy, and expensive. But if the company could meet the specs for Ginger's gyros and do it in volume, on Ginger's schedule, at a cheap price, that would solve a costly problem. BAE Systems, the

British aerospace engineering firm, had reduced its price per gyro by $10, but each Ginger needed five of them. "So I'm going to call J. T.," said Dean, "and say, 'I have an unreasonable request that might turn into a nice piece of business for you.'"

The investors' meeting started early the next morning in the War Room. Tim Adams's new assistant had upgraded from the usual cardboard box of donuts to a fancy spread of muffins and Danish nestled inside a basket lined with checked cloth. She had surrounded it with decorative squash, pumpkins, and Indian corn. Tim walked in and stopped short. "Whoa!" he said. "Look at that! Rachel doesn't know the drill yet. We need the day-old stuff."

Michael Schmertzler was there. Aileen Lee, on the speakerphone, represented John Doerr.

Mike Ferry reported on the Boston Riders Panel. Everyone, men and women, had felt confident within fifteen minutes. About half of the thirty-six riders understood Ginger's broad applicability and didn't consider it merely recreational. But some riders had also said that their commutes to work were too long for Ginger, or that they preferred to walk for short errands. This last comment troubled Mike.

Brian Toohey was wearing a cast. Schmertzler asked how he had broken his arm. "It's not his arm," said Dean, "it's just a little bone in his wrist."

"Who has a broken arm?" said Aileen Lee over the speakerphone.

"Our chief safety officer," said Schmertzler dryly.

The man who had been hired to handle commercial sales reported that he intended to pursue "trophy accounts" first: UPS, Amazon, Chrysler, General Electric. He wanted to "enter high" by going straight to the CEOs and boardrooms, using Dean's, Doerr's, and Schmertzler's contacts.

Doug briefed them about engineering progress. Schmertzler asked about range and batteries. The target was seventeen miles, but they were getting only half that. Schmertzler looked startled. Doug said he wasn't concerned. The motors were fighting each other, and the team would fix it, especially if they ever got some software engineers, his number one issue. "Is this on the critical path?" said Schmertzler, referring to the lack of software support.

"It *is* the critical path," said Tim Adams. "It's the gating item. We have a recruiter who specializes in getting people who aren't looking for a job. He checked 109 companies, found 120 candidates, and delivered *two* résumés." Tim mentioned the $5,000 signing bonuses.

"So the whole project is getting backed up because we need three software engineers?" asked Schmertzler.

"Yes," said Tim.

Schmertzler looked appalled. "If it takes $5,000 to blow someone out of another job, we should do it," he said, "because we have tens of *millions* of dollars backed up behind that." That had been Tim's argument to Dean for months, but Tim and Doug had been shackled by Dean's refusal to pay competitive salaries, much less bonuses. Schmertzler didn't know that, and was blaming Tim for being shortsighted. Dean let Tim take the blow.

Scott Frock, the financial manager, began his report. Schmertzler interrupted to warn that Acros now had a huge bankroll, nearly $90 million, and it was human nature to spend it. He was concerned about preserving capital. Tim said his forecast for cash on hand at launch had dropped by $6 million, though he hoped to improve that.

Schmertzler looked unpleasantly surprised and asked how much money had gone out between April 1, the starting date of the financing, and September 30. About $12.3 million, said Frock. So they expected to spend another *$66 million* before launch? asked Schmertzler. From Schmertzler's perspective, this looked like another blunder. First Ginger's managers balk at spending a few paltry thousands to attract crucial software engineers, then they budget millions sloppily. Penny wise, pound foolish.

"Guys," he said sharply, "just so there's no misunderstanding—at the July board meeting, when the number for sales and marketing rose, no one said anything, but eyebrows did rise." And now the number had jumped again, to almost $5 million more than Bob Tuttle's initial estimate. "Those are important numbers, guys," he said, his voice edged. His associate Dan Clare would visit soon, he added, to drill into those figures and determine why they had ballooned.

"I agree," said Tim. "I need to know how some of these numbers grew." Schmertzler snorted, implying that Tim should already have the answers, just as he should have spent the pittances for software bonuses.

Schmertzler studied the spending sheet again. Doug's product development costs had nearly doubled. The nine-month slip explained much of that. But the slip shouldn't affect marketing's budget, so why had it puffed out by nearly $4.5 million?

"To be honest," said Mike Ferry, "is there any fat in that? Maybe a little." Not much of a defense. And what had Tim been doing while Mike Ferry's budget fattened? Schmertzler pierced Mike with a stare and then dove back into the numbers. The table stayed silent as Schmertzler concentrated, marking some figures, briskly circling others. It clearly rankled him that all these numbers were still in volatile flux, still unsettled and undependable.

After a minute or two, he looked up. "For presentation purposes," he said, referring to the upcoming board meeting, "it's *shocking*"—he paused, evidently shocked by his own vehemence—"or rather *surprising*," he continued, "to see movement from $27 million to $52 million" between the original April 1999 plan and the current plan. He wanted Tim, Dean, Bob Tuttle, and Scott Frock to get together immediately and press the team's managers about their numbers.

Tim Adams had seemed ill informed and knew it. Afterward he told me that he felt terrible because he should have known all the budget breakouts and had been asking Scott Frock for them. But Frock, a relatively new hire, had been so busy getting the financial side of the business up and running that he hadn't had time to pull them out. Tim hadn't shifted the blame to Frock in the meeting and consequently had looked bad in front of Schmertzler. So had Mike Ferry.

As companies have come to rely on information systems to integrate data, e-mail, and the Internet, the position of chief information officer (CIO) has become critical. By the end of 2000, the dot-com madness had inflated things to the point that some CIOs were demanding seven-figure salaries and 2 percent of the company. That wasn't going to happen at Dean's place. To save money, Dean and Tim were considering hiring a retired CIO who would come in twice a week to oversee lower-level IT people.

One evening in early November, Dean, Tim, Doug, Scott Frock, and Brian Toohey interviewed a promising candidate in DEKA's conference room. Patrick J. Zilvitis had worked at IBM, PerkinElmer, Martin Marietta, Digital Equipment, and Gillette. He had retired from Gillette a few months ago as CIO and was looking for something to occupy him a couple of days a week.

As he recounted his work history, Dean kept asking what he meant when he said "*We* did this and that." Zilvitis stayed vague. "What I need to know," interrupted Dean, "is did *you* do these things or was it the mouse in your pocket? Because if it was the mouse, that's who we need, not you."

After a pause Zilvitis said, with a bit of an edge, "Fair enough," and launched an impressive list of activities, each one beginning "I was personally involved in . . ." Sharp and savvy, Zilvitis handled questions with a casual depth of expertise.

When he left the room so the group could discuss him, they quickly agreed that he would be a catch. "Make sure he doesn't want a substantial piece of the company," said Dean to Tim. "If it's reasonable, we'll do a confidentiality agreement and show him Ginger and go to dinner. We may get a very quick no. It wouldn't be the first time."

The negotiations about salary took longer than expected, or at least longer than Zilvitis expected, and weren't resolved. But Dean decided to clear him and show the Fred and Ginger videos as bait. Introducing them, Dean said that many of the first Freds and Gingers had been built at his house. "But I'm far from engineering now," he quipped. "It's why I'm manically depressed."

Then he bombarded Zilvitis with the spiel: pollution, choked inner cities, Third World needs, 1.1 million postal carriers, 13.6 million college students. "You worked for IBM," he said, "and they never saw the PC coming, which was going to wipe them out. And the PT—the personal transporter—will do the same to the auto industry."

Twenty minutes later, with Zilvitis looking restive, Dean finally popped in the Ginger tape, and the Talking Heads started singing, "We're on the road to nowhere." Dean provided a running commentary. At dinner he continued to sell Zilvitis without pause. Acros expected to ship between 200,000 and 500,000 Gingers in the first year. Zilvitis stayed rather quiet and seemed unenthused. By the time the evening ended, I didn't expect him to jump in.

But a few days later, he did. Ginger had itself a 22-carat half-time CIO. A bargain.

The second Ginger board meeting commenced in DEKA's conference room on November 13, a gray New Hampshire day. In addition to Dean, Tim, and Bob Tuttle, the participants included Michael Schmertzler, Paul Allaire, and Vern Loucks. Aileen Lee was on the speakerphone again for John Doerr. Only Allaire wore a tie.

Tim started with the bad news: They were behind on D3 because they needed software engineers. Other things were progressing. Parts were freezing and the information technologies (IT) contract was coming along—Oracle wanted $2 million to build Ginger's network and $200,000 per year to manage it. Tim suspected that Doerr had talked to Larry Ellison, Oracle's founder and CEO, because Oracle had sent its A-team. Tim had started interviewing public relations firms, he said, "in case we have leaks and need to put fires out." Burson-Marsteller, the leading PR candidate, also had sent its A-team—its crisis guy had worked on Tylenol, New Coke, and Bhopal. "These guys *love* trouble," cracked Loucks. A possible problem: One of the firm's big clients was Ford. As for Ginger's cost of goods (the cost of basic materials), they were $350 over the target, but Tim expected to meet it within a year of launch.

Mike Ferry introduced the new company name: Flywheel. The logo was a stylized backwards F, with a rounded bottom, topped by a dot—an abstraction of a moving person on wheels. It was intended to look dynamic without conveying dangerous speed. Mike hoped to buy the domain name flywheel.com from its current owner, and had pre-emptively registered troublesome names such as flywheelsucks.com.

Mike presented the good news about the Boston Riders Panel—the quick learning curve, the enthusiasm, the safety. He added that people with technical knowledge expected it to be expensive, but non-technical people—that is, most people—saw a simple, presumably inexpensive machine. That was worrisome, as was the finding that fewer than a quarter of the riders said they probably would buy a Ginger. Mike hoped to start "group interactive" consumer research by March 2001 to start collecting real data.

Don Manvel, director of operations, reported that since July's board meeting he had hired three procurement people and four plant people. They were looking at yet another potential site for the plant and hoped to nail down a deal within a week.

Brian Toohey spoke about regulatory issues. He had removed the cast from his wrist so he wouldn't have to answer embarrassing questions. He confidently declared that the regulatory risk wasn't great, as long as the company involved government officials early at all levels. He wanted to start showing Ginger to such people during the coming year, but warned that leaks could occur. Though twenty-eight states restricted motorized vehicles on sidewalks, he would argue that current regulations didn't apply to Ginger. If that didn't fly, he would work to revise local laws. One house of the New Jersey Assembly had recently passed a bill restricting skateboards and motorized scooters from roads and sidewalks, but the state senate and the governor hadn't yet taken a position. Brian had been working to insert a definition into the senate version that would exempt Ginger. He predicted that European governments and cities would welcome the machine. His impressive presentation soothed some of the board's worries.

Some of the budget numbers that had concerned Schmertzler had been scrubbed. The marketing budget was lighter by $3 million. As of the end of the third quarter, the project had spent $11.2 million. By the fourth quarter of next year, just before launch, Scott Frock expected the burn rate to be $12 million to $13 million.

The question Schmertzler posed was whether the company could make money on a machine priced at a few thousand dollars.

Bob Tuttle had estimated the margins for the Metro and Pro, and they were disturbingly low.

"I'll be right back," said Schmertzler, suddenly standing. He left the room.

"He went to blow lunch," said Dean.

When he returned, Tim asked what issues the board wanted him to focus on. "Cost of goods," said Dean quickly. Loucks voted for regulatory and international distribution. Schmertzler wanted a tighter, more accurate financial model and a marketing strategy sharpened to the point of naming targets. Tuttle wanted a senior appointment in sales.

Tim nodded at each request as reasonable, then said that the team was walking a tightrope. He didn't add that Dean had forced them onto it. "It's hard to know if you have the right marketing targets if you can't talk to people," said Tim. "It's difficult to know if your regulatory strategy is correct if you can't talk to people."

"Dean, it's really something you need to think about," said Loucks, going straight to the cause.

"I agree," said Schmertzler. "I continue to think a sexy leak could be beneficial."

"You need a reveal strategy before you can have a marketing strategy," said Loucks. He suggested that they design one by February, just three months away.

"Do you agree with that, Dean?" asked Schmertzler.

Dean looked cornered and unhappy. "Yeah," he finally said in a low, tight voice. Completely unconvincing.

chapter fourteen

Everything Is Connected
All the Time

November–December 1, 2000

O n November 14, the day after the board meeting,
Michael Schmertzler's genial young associate, Dan Clare,
spent the morning going over figures with Ginger's
financial director, Scott Frock. To attract mass consumer sales, Dean, Tim Adams, the marketers, and even
outsiders such as Jeff Bezos considered a few thousand
dollars to be the machine's upper price limit. On
Schmertzler's instruction, Clare had been trying to discover what the profit margins would be at various price
points. The figures he found weren't thrilling to this
investor of $38 million. Schmertzler wanted a margin
four or five times higher. Maybe they should launch
in the commercial market first, Clare said, where they
could charge businesses considerably more and make
some real money that would help fund a big consumer
launch later.

Clare also showed Frock some penetration curves
for disruptive new products. Typically the graph line for
innovations such as the PC and the VCR stayed low for

many years before tilting up. DVDs climbed the fastest—the graph line headed northeast after only two years. By comparison, Bob Tuttle's projected penetration curve for Ginger lifted off like a Harrier jet from an aircraft carrier.

Schmertzler and Clare also worried that the Ginger team might blow through a lot of cash and run dry in the middle of the consumer launch, before revenues could catch up to spending. Venture capitalists encouraged new companies to spend money fast, because then the start-ups needed to borrow more, at less favorable terms. Clare told me that Credit Suisse disliked the VC model—he didn't mention the elephant in the room, Kleiner Perkins—and preferred to ensure Ginger's long-term health by reining in expenses now.

Clare's worries echoed an article from the front page of the previous day's *Wall Street Journal*. The story described the swift rise of ICG Communications, Inc., whose clients included Microsoft, Earthlink, and NetZero. Its headquarters alone had cost $33.5 million. The company had grown too fast and borrowed heavily to do so, a cycle that left it cash-poor. The market responded by dropping ICG's share price from $39.25 eight months ago to 44 cents.

"But its fatal error," said the story, "was to promise super growth and then disappoint its investors. As the go-go capital markets of the 1990s recede, fast-growing companies with huge debts but slim or no earnings are extremely vulnerable."

Tim Adams had been worrying about budget overruns, too. He still hoped to have $20 million in the bank on the first day of launch in late January 2002, but it wouldn't be easy. The accelerated schedule worried him as well. How could he meet it if he couldn't even find job candidates, much less recruit them?

And then there was the factory—they should have settled a deal a year ago. The plan called for a plant to be finished by the end of June 2001 so that the V1s, the assembly-line test model after D3, could be built there in August. That possibility had evaporated. Dean and Bob Tuttle *still* hadn't decided anything. Dean would argue that putting off the decision had saved money. Tim had argued, without effect, that the delay endangered the launch. Dean and Tuttle were now considering four more sites. The difference between them amounted to between

$20,000 and $50,000 a year, whereas slipping the schedule a month could cost $10 million. "So pick one and *move on,*" said Tim. "But Dean and Bob will work it until they've squeezed out every nickel." Like the reluctance to pay software bonuses, this parsimony was hurting the project. Dean and Tuttle disregarded advice that Tim knew was critical. It was getting more and more frustrating.

Dean's mode of operation puzzled Tim, who agreed with Ron Reich that making quick decisions and absorbing mistakes put you farther ahead in the long run. Tim was still riding that DEKA learning curve. He often felt like he was doing it with his hands tied. Dean, Tuttle, and Schmertzler were clamoring for a distribution plan, yet Dean also insisted on absolute secrecy. The two demands were incompatible. They also wanted a detailed marketing plan, but that usually came *after* the distribution plan, which smacked them into the wall of secrecy again. Dean sometimes acted irritated that he was paying three marketers who wouldn't be able to do a marketing plan for months. He didn't seem to realize that the marketers had to be on board early, for many reasons. Launching a new product category required lots of preliminary strategizing, and the product needed a name and a logo, especially for dealing with suppliers and distributors. Those were marketing tasks, which meant paying marketing salaries and expenses. Such things were obvious to Tim, but he didn't think Dean or Tuttle understood.

Tim also felt tied up in other ways. If he acted like a strong leader and tried to run things, Dean resented it and felt usurped. But Tim knew that if he surrendered to that, he would paralyze himself and lose the team's respect. He already hesitated to stand in front of people and declare the company's direction. He hesitated to recruit people because Dean had to approve every offer. He hesitated over decisions because Dean second-guessed every detail. He compensated as best he could, giving Dean warnings and advice, but living under Dean's thumb was getting more and more frustrating. Tim kept telling himself that things would be different once Ginger took off and Dean turned his attention elsewhere.

Mike Ferry loved JDK's logo and typeface for Flywheel. That stuff was the fun part of his job. Still, it took skill and experience to manage such firms and extract their best work. "Dean doesn't understand that, and

Dean doesn't value that," said Mike, with a trace of bitterness. "He thinks it's all nonsense."

Mike felt certain that hiring outside firms was the way to get top-notch work, but Dean hated paying consultants and preferred to hire people. It was cheaper and gave him more control.

Mike's associates, Tobe Cohen and Matt Samson, were on their way to check out COMDEX, an annual three-ring circus of technology in Las Vegas. Mike considered it a possible launch site for Ginger late next year. He said not to mention the trip to Dean, because spending money for travel was such an issue with him. He paused, then added that Dean probably had complained that the marketers' trip to Japan was wasteful. Mike always seemed placid, but now irritation crossed his face. "If you want to launch somewhere, you can't just read a book," he said. "You have to go there and check it out. Dean doesn't believe that."

For the second time in a year, the Ginger team, which now numbered almost fifty, had outgrown its quarters. Builders were renovating large spaces across the hall and downstairs across from the lab. J. D. Heinzmann wanted to name the eleven new conference rooms after great inventors. Someone else suggested preserving the Ginger theme by using the names of dances. An engineer lobbied for international cities, to foreshadow a global company, and the team chose that theme. From now on meetings would be held in Tokyo, Bombay, Rome, Amsterdam, London, Paris, San Francisco, New York, Hong Kong, and Kuala Lumpur. And in Ulaangom (Mongolia), a nod to Ginger's beginnings in the Penthouse, where the engineers had labeled a wall clock with that obscure town's name. Don Manvel thought they should run the names by Dean, but that was vetoed. "Let's not involve Dean in things he doesn't need to know about and get bogged down with his opinion," said an engineer.

As winter approached, the atmosphere around Ginger seemed quieter. Less joking around, less raucous laughter, fewer congregations of kibitzers. It was partly that the team had split up. Phil LeMay, J. D. Heinzmann, John Morrell, and the rest of the vehicle dynamics group had moved downstairs. People also were busier, traveling to visit suppliers and focusing on smaller tasks that rarely involved the entire team.

Doug missed having his whole team in one place, sparking and correcting each other.

The new space required furniture, which required expense, which Dean hated. Tim and the team leaders met several times to discuss the arrangements. Doug dreamed about open space with a playful, harmonious aesthetic and whimsical touches like conference tables made from airplane wings.

Then they met with Dean's property manager to discuss options. The most economical plan, he told them, would be to use some old 5-foot partitions now in storage, "kind of a brown color." As for consistency of appearance, that might be possible in certain areas, but not overall. And as for quality, "You don't want the guys two buildings over"—at DEKA—"thinking you're the golden cow. Dean would have a riot on his hands." His design vocabulary didn't include airplane wings.

Tim, Doug, and Mike disliked these choices and put off the decision once again. The discussion seemed to take place on an infinity loop. They couldn't make a decision because they worried about Dean's reaction to the bill.

One day in mid-November Dean's assistant pulled him out of a meeting for a call that sounded important. This is the Office of the President, said the person on the phone.

The president of what? asked Dean.

Of the United States, answered a White House staffer. You've won the National Medal of Technology.

Dean seemed pleased, especially if he could milk it for publicity for FIRST. A few days after the call, he asked me to help him revise the executive summary that he was sending to Washington about his work. We were sitting in his library.

As he complained about how fast the Ginger team was spending money, he looked up the day's price for Paul Allaire's Xerox stock. He winced. "It's in the eights." He looked up Amazon and winced again. "It's $28 and change, down from $113. Jeff Bezos has lost several *billion* dollars." Things weren't going well for Steve Jobs, either. Apple had dropped from $60 a share to the low $20s.

Dean's in-box held a fresh e-mail from John Doerr, addressed to Jobs and copied to Jeff Bezos and Dean. Doerr wanted this group to meet on December 8 in California. Dean checked his PDA and said, "I can do it." I wondered why Doerr wanted the meeting. It sounded like a West Coast coup.

"Exactly," said Dean, grinning. "I think Doerr still wants them involved."

"It's turning into another pissing match between Doerr and Schmertz-ler," said Dean two weeks later. That morning Doerr had e-mailed Dean that he wanted to bring two more people from Kleiner Perkins to the San Francisco meeting. Schmertzler had hit the roof. He wasn't going to travel all the way to the West Coast for a short meeting— scheduled for an hour and a half—that was sounding more and more like "a John Doerr love fest." Now Dean was fretting that if he went, he would annoy Schmertzler. If he didn't, he would annoy Doerr.

Meanwhile, Dean added, the stock prices for a lot of Kleiner Perkins's companies weren't just down, they were disappearing. Some-one had sent him an article from *Fortune* that quoted Doerr at the In-ternet Summit, described by the magazine as "the biggest recent gath-ering of Internet superstars." Doerr had told the dot-commers that he wanted to get back to basics and invest in "atoms, not bits." "I'm inter-ested in making things again," Doerr had said. "Clean water, trans-portation, clean power—those are the big markets of the future."

"He might as well have said 'the DEKA Stirling engine and the Flywheel,'" said Dean. Maybe the Old Economy was the new New Economy.

In keeping with the increasing emphasis on sidewalks and safety, two lawyers from Washington, D.C., visited Ginger to give the team a scare talk. One specialized in product liability, the other in litigation. But first Doug introduced them to the machine. Dressed in an untucked shirt, jeans, and flashy running shoes, Doug looked like a grad student. De-spite a heavy cold, he gave an entertaining performance—low-key, clever, fluid.

He told the lawyers that Ginger had no brake or throttle. Like the

F-16, it was a fly-by-wire device. Most steering mechanisms were connected to the wheel by rods or other mechanical means, but in Ginger, electrons provided the connection. Would the liability lawyers be interested in seeing a few potential failures? Doug stepped onto Ginger, then pulled the "kill switch." The machine instantly lost power and fell over as Doug jumped off. He simulated a short: The machine spun like a dervish. He explained what it meant to run out of headroom. All of these possibilities, he said, were being designed out.

Doug asked one of the lawyers to climb aboard. The product liability guy got on but stood so stiffly he barely moved. Doug told him to lean forward and speed up, so he did, then panicked and crashed into the lab wall, leaving a black skid.

"Excellent," said Doug, grinning. "Now we can have a productive discussion."

He showed them Ginger's innards. There were redundant motor control boards, meaning that if one board or battery pack or motor failed, the other side took over. Another board lived in the dash. All three boards caucused constantly, and if one disagreed, the other two could outvote it and bring the machine to a stop. In the inertial module unit (the House of Gyros), at least two gyros monitored motion in any axis.

Doug told them that Ginger was engineered to minimize failures that caused bodily harm, and hence lawsuits. For instance, if the control shaft ever came out, the rider would go down, so the design required the shaft to be fully seated before the electronics would connect. Next, two bolts secured the shaft to the chassis. If the bolts loosened and the shaft started to wiggle, the pins in the connector were different lengths for the different computers, so one side would shake loose before the other, and the good side would tell the machine to slow down and stop. But of course it wasn't possible to engineer-out foolishness, because fools were so ingenious. People were sure to sit on the machine or stand on the front edge or tote someone.

Doug took the lawyers through Mike Martin's shop of horrors. A D2 had just completed a trip of 5,200 miles on the roller machine, and still ran. Mike had recently dropped another Ginger 3 feet onto a pipe, then ridden it away. Next he had tilted it and dropped it. That fall bent one of the boards and drove the batteries through the chassis, but the machine still ran.

Doug moved on to the row of Ginger's competitors parked in a far corner of the lab. He identified the flaws and dangers in half a dozen scooters and electric bikes. Liability lawyers could live handsomely off them. Pointing to a popular scooter, Doug told the lawyers that one of Ginger's engineers had tested it by riding it home in the rain and then plugging it in to recharge. The scooter had suddenly revved, spurted away, and knocked down a table. Doug gestured at another motorized scooter. "Here's my favorite warning label," he said, reading its ungrammatical phrasing aloud: "WARNING: Do not pull inward on the yellow emergency lever during operating the vehicle, only when accelerator is out of control."

"Not *if*," noted Doug, grinning, "*when*." His team had eliminated the whens and was hunting down the ifs.

Engineers spend a lot of time staring at drawings, sometimes on paper, more often on a computer screen. They look for possibilities they've missed or escapes from realities they've banged into. They stare for a while, doodle with a pencil or a mouse, stare some more. On some days, a time-lapse video of an engineer at work would boil down to this: stare, doodle, stare, sigh, stretch, caffeinate, trade quips with colleagues, stare, caffeinate, doodle, show drawing to colleagues, stare with colleagues, trade quips, stare, sigh, go home.

Today a group of engineers and industrial designers were standing at a desk, studying some rough drawings of a small section on the dashboard. They were figuring out, or not figuring out, how to hide the stem bolt with part of the dash module, yet also ensure that the bolt would be accessible for servicing. An adhesive would work, but then reaching the stem bolt would require destroying the part glued over it.

"You've got a lot going on in a very small space," said Kevin Webber, a design engineer, to Pat Kelly, a young engineer working on the problem. "Are you worried about that?"

"Yeah," said Pat in a soft low voice. "I'm worried."

"Good," said Kevin brightly, patting him on the shoulder. "My job is done here." He walked away chuckling.

A design bump that looked tiny often turned out to be a 10-foot scaling wall, usually with a pile of bashed-up engineers and an industrial designer or two crumpled at the base. "The devil is definitely in the details," Pat said later. "We originally had six bolts and got it down to one, so we said, 'We'll just cover it up.'" The task sounded so simple, he started to laugh.

A few days later Doug sat down in one of the new conference rooms with Pat, Bill Arling, and Scott Waters to talk about the dashboard module. They hoped to freeze as much of it this afternoon as possible, including the new handlebar, the shape of the iButton, and the grip measurements. An early winter dusk was darkening the room.

They worked their way down the list, freezing things. After fifteen minutes Doug said, "From here on, they're all contentious." They reached the stem bolt. Pat and Scott were still looking for a way to cover it to discourage tampering, but without prohibiting access. Scott explained that he wanted to link the iButton visually to the LCD battery display, so neither was isolated. An elastomer could accomplish that and plug the stem. They couldn't put the LCD right on the surface because it would get scratched easily, but they couldn't lower the surface of the elastomer enough to sink the LCD because of the electronics board beneath it.

"How long have you spent on this detail?" asked Doug.

"All yesterday," said Pat.

"When does it have to be released?" said Doug.

"Tomorrow," said Bill Arling.

Scott did a double take. "Are you *serious*?" he said.

Bill nodded. To give the manufacturer the necessary lead time, the part had to be released on December 1. Today was November 30. Scott looked twitchy.

"I'm going to do something you won't like," said Doug. "I want this frozen by 12:00 tomorrow. We can't spend any more time on this detail when there are structural and functional issues to deal with."

"OK," said Scott, "but if this can't change after D3, I need someone to stay tonight and help Pat and me, because otherwise we might end up with something we're not happy about."

"How much is the tooling?" said Doug.

It was more than $200,000. In other words, changes weren't likely after D3.

Scott had moved from twitchy to tense. "We're so close, I'd hate to give up," he said.

"You don't have to," said Doug, "until tomorrow at noon."

"We're *so close,*" Scott repeated. He sounded in pain. He wanted this detail to be perfect.

Doug considered him. "Would giving you until Monday really make a difference?" he said. It was Thursday.

"*Yes,*" said Scott. "If I had the weekend—." He looked at Pat. "If you can do that."

"Sure," said Pat.

Doug turned to Bill. "Does that take Pat from other tasks?" Now Pat looked uncomfortable, squeezed between hard places. "Are we behind?" persisted Doug.

"Yes," said Pat.

Doug looked from one to the other. "You have to keep your priorities straight," he said. "If you spend four days on this detail and you're already behind, it might mean you spend *half* a day on something really important."

"We're *so* close," said Scott.

Doug considered him again. "How many hours?" he said. "Not days. Hours."

"I think it *is* hours," said Scott.

Doug paused. "OK," he said. "But I want an e-mail as soon as you have it, with a call on my cell phone."

Scott looked relieved, until someone made a crack about his phallic design for the elastomer. Bill and Pat had been ribbing him about it. It seemed to be a Rorschach test for engineers.

Scott looked mortified. "That *doesn't* leave this room," he said. "I don't want to bias anyone else before they see it. OK? *OK*?" The engineers just chuckled.

J. D. Heinzmann disagreed with the diagnosis that Ginger was getting only half of its expected seventeen-mile range because the two motors were fighting each other. He thought they just weren't sharing equally, an entirely different dilemma. One piggish motor was demanding extra juice from the batteries that fed it. The pig could get away with this because the power system was using voltage mode, which not only permitted the disparity but institutionalized it by always pulling the same unequal voltage. So if the amps asked for 50 volts, the pig might always get 51 while its deprived twin got 49. Voltage mode proved that consistency could be a flaw. It allowed the greedy motor to drain its battery pack faster, thus shortening the machine's range.

Instead of using voltage mode, J. D. wanted to mimic duty-cycle mode. When duty-cycle mode detected one battery pack grabbing more than another, it intervened to enforce equality and cooperation. "It takes from the rich and gives to the poor," said J. D.

But why did he want to *mimic* duty-cycle mode instead of using the real thing? Because, he explained, the power bridge within the amplifier introduced distortion into the duty cycle, which had to be circumvented with an imitative algorithm. The gist of this techno-speak was that he was tricking the technology, jiggering it, right? J. D. grinned and nodded.

Most civilians assume that two matching motors or batteries or amplifiers will perform identically. But that's never true. Mechanical equality is always artificial, achieved only through compensatory engineering that tricks parts into acting the same way.

Earlier in the process, when the goal wasn't precision but mere operation, the jiggering occurred at the rough end of the scale, with electrical tape, twisted wire, and plywood. Phil LeMay had condensed this stage into a maxim: "Measure with a micrometer, mark with chalk, cut with an axe." At the beginning, engineers disregarded precision in the rush to make something work.

"In engineering school everything is idealized," said J. D., "and then you learn that reality isn't like that. Things are nonlinear. There's friction. We ignore those things at first because it's easier to calculate without considering them. And then you get out in the world and have experiences and realize that nothing is perfect, that you have to compensate for all sorts of things in all sorts of ways."

This struck me as rich metaphorical territory, but for J. D. it was real, with technical consequences and verifiable solutions. A compensatory twitch in one part of the machine could cause chaos in other parts, like the so-called butterfly effect, which holds that tiny wings fluttered in California help cause blustery weather in Maine. Everything is connected.

"Yes," said J. D., nodding vigorously. "Everything is connected *all the time.*"

That described DEKA as well, with Dean as the butterfly generating weather systems from his corner office. Sometimes the conditions were turbulent. Part of the financial deal that created Ginger as a separate company stipulated that any new intellectual property developed solely by the Ginger team would belong to the new company. But two of Dean's bedrock principles were never to relinquish intellectual property rights and never to give up control.

Before the financing came in and Ginger officially became a separate company, Dean had decided to retain some of Ginger's engineers for DEKA while letting them continue to work on the project. This would give Dean the power to move the specified engineers back to DEKA whenever he chose. It was also another excuse to break down the wall that he perceived going up between Ginger and DEKA. It was understood on the Ginger team that this arrangement of "loaning" DEKA employees to the project also would make it unlikely that any technical breakthroughs or patents could be claimed solely for the new company.

John Morrell, the lead controls engineer, was among those at Ginger most likely to design breakthroughs, and the rumor was that Dean intended to keep him on the DEKA list. John objected strongly. He wanted to be part of the new company, not DEKA, where Dean would have more control over him financially and professionally. Dean shrugged off his objections and didn't back down until John threatened to quit and seemed to mean it. Losing him would have crippled the project.

When the list of DEKA retainees finally appeared, it included four members of the Ginger team: J. D. Heinzmann, Mike Martin (in charge of mechanical integrity), Jon Pompa (a valuable utility engineer who could solve problems across disciplines), and Mike Slate, the team's indispensable mechanic. Mike Martin thought there had been a mistake and asked to be moved to the Ginger list. Dean responded by calling the four retainees together and instructing them that *he* would decide where they belonged and what work they did.

John Morrell wasn't the only member of the team who considered quitting because of Dean's autocratic style. Don Manvel, Ginger's director of operations, almost left several times because Dean made his job so difficult by his relentless manipulations of suppliers in an attempt to get a better deal. Even Doug Field, Ginger's main structural beam, considered quitting when Dean seemed about to yank him out of Ginger to straighten out the iBOT project, and when he blamed him for the slip.

Speaking about Dean's strategy of retaining the four Ginger guys for DEKA, a member of the team said, "It partly had to do with Dean feeling like he was losing control, and he doesn't like that. We knew there would be extreme separation anxiety, but we didn't know what form it would take." It was the downside of Dean's usually benevolent patriarchy.

The potentially divisive separation of the team into Ginger employees and DEKA employees was common knowledge, but the only one of the four who ever spoke to me about it was J. D., indirectly, a few minutes after explaining that nothing is perfect and everything is connected. Because he was on the DEKA list, he said he sometimes felt like he was working at Ginger on a visa. Dean had told him that he kept him for DEKA because J. D. surely would want to return there someday to develop new products.

"But I've realized that the most important thing to me isn't the work, it's the people," said J. D., "I really like the people here at Ginger. When I was young, what excited me was the technology and the project. Now I want to be on a good team, with good leadership, with people who are fun to work with. Dean doesn't understand that."

Because the technology and a cool new project were what mattered most to Dean? "Exactly," said J. D. "And that's the way he runs DEKA. So my worst fear is that he'll move me, but I'm trying not to focus on that anymore."

At the end of November, Tim held another "town hall" meeting in the lab to update everyone on developments. They had finally signed a purchase agreement for a plant site in Bedford. Dean was buying the land and leasing it to Ginger. The plant would have about 75,000 square feet, with ten or twelve loading docks, and would employ about one hundred and fifty assemblers who would make two thousand Gingers per day.

Doug Field noted that his team still had "a gaping bloody hole" in software, so the D3 schedule continued to slip. He wanted to halt that wherever possible. "We can't succumb to the temptation of waiting just a *little* longer to get it right," he said. They needed to move on. But he also wanted them to pace themselves. "We're at that point in the marathon called Heartbreak Hill," he said, with most of the race behind them but some distance to go. "I don't want you to get to launch exhausted, because the goal isn't just to build a product, but to build a team and a company. So sometimes you need to step back and get some perspective. I would ask that you don't change your standards, but also don't steal from your families or from our product." That sounded like a balance of forces as complicated as Ginger's.

In his clipped way, Ron Reich asked, "When. Do. We. Become. Flywheel?"

"Not. For. A. Long. Time," said Tim. People chuckled. "We're still stealth, so don't refer to Flywheel with anyone."

But Dean had waited too long to contact the owner of the domain name, who had sold it thirty days earlier.

The next evening in Washington, D.C., Dean received the National Medal of Technology. The other awardees included Douglas C. Engelbart, father of the computer mouse, hypertext, and other breakthroughs in personal computing; three men from Corning Glass credited with instigating the telecommunications industry by inventing low-loss fiber-optic cable; and IBM for innovations in hard-disk technology and storage products. These were the sort of hard-tech innovations that Dean could respect, even if they had led to what he often called "dot-com nonsense."

Dean's dark pompadour contrasted with the grays and whites on the dais. He had submitted to the gravity of the occasion by wearing a tux, an extraordinary change of costume. He seemed to be stretching his neck to escape the stiff collar.

In her opening remarks, Undersecretary of Commerce Cheryl Shavers said that the award winners were not simply innovators. "They challenge the status quo," she said. "They not only ask why, as scientists and researchers—they ask why not?" That certainly described Dean. So did the remarks by Secretary of Commerce Norman Mineta, who praised the laureates because they "pursue seemingly impossible dreams," and catch them "through extraordinary vision and persistence." Mineta placed the laureates in the same company as Edison, Bell, Ford, and

Marconi, "whose accomplishments forever changed the world." The laureates, said Mineta, "exemplify the American spirit—vision, ingenuity, hard work, and the courage to be different, the courage to think beyond the status quo."

Shavers read the citation for Dean: "For inventions that have advanced medical care worldwide, and for innovation and imaginative leadership in awakening America to the excitement of science and technology." As the applause swelled, Mineta draped the medal over Dean's bowed head. When Dean straightened up he flashed a big grin.

Maybe he was thinking about his plans for the White House. The day after the ceremony, President Bill Clinton was holding a five-minute photo op to congratulate the medal winners. Dean, as usual, had his own agenda. He wanted to take along the iBOT, some FIRST material, and a crew from *60 Minutes II,* which was doing a story on him. The White House staff told him no, *no,* and *No!*

So he asked mutual acquaintances to call Clinton on his behalf. These flanking maneuvers irritated the Clinton staffers, one of whom called and told Dean he could *not* have more than five minutes, could *not* engage the president in any discussion or pass him any literature, and abso*lutely* could not bring a 200-pound machine into the Oval Office. And as for Dean's request that the crew of *60 Minutes II* be allowed to accompany him to the Oval Office, *forget it.*

The *60 Minutes II* people happened to be in Dean's hotel room when he took the call. They asked what the verdict was from the White House. Dean said it looked like a definite maybe, and to come with him tomorrow.

The next day Dean rolled up to the White House gate on an iBOT, trailed by the *60 Minutes II* producer and cameraman. He charmed the guards, letting them think he was disabled and had permission for the others. The guards did make him empty the pockets of his fatigue jacket and explain why he was carrying not just a cell phone, camera, and Palm Pilot, but a tape measure, screwdriver, flashlight, electronic distance calculator, collapsible telescope, electrical tape, voltage meter, adjustable wrench, velocity meter, and, most puzzling of all, a steel machine nut as big as a baseball.

He and the TV crew got through the gate and proceeded up the path toward the White House, Dean speeding along balanced on two wheels. Then he hit a thick TV cable and the iBOT started to tilt forward, which triggered the computers to transition the seat downward and stop the machine. Dean jumped off. "I was showing off," said Dean, recounting it. "It was so stupid, because then the chair has to be reset with a computer, and here I am halfway to the White House."

He called Benge Ambrogi, who had already pulled away. By the time Benge returned, Dean had dragged the iBOT back to the gate. "You didn't tell us you could walk," said the guards. "You didn't ask," said Dean.

He finally wheeled into the Oval Office on the iBOT, trailed by *60 Minutes II,* and proselytized Clinton for fifteen minutes about FIRST. Clinton had met Dean before at a Rose Garden ceremony honoring the FIRST champions. "If you do not want to hear about what he does," Clinton remarked at the time, "*do not ask* or stand within a four-mile radius." When the photographer started shooting, Dean whipped a folder of FIRST propaganda from beneath the iBOT's seat and held it in front of the president.

And so Dean got everything he wanted into the photo—the president, the iBOT, the folder labeled FIRST, plus some footage of Clinton nodding about FIRST for *60 Minutes II.* No wonder Dean was smiling so widely as he shook Clinton's hand.

At the next major stop on his itinerary, it might not be so easy to manipulate things. He was headed to San Francisco to face John Doerr, Steve Jobs, and Jeff Bezos.

chapter fifteen

West Coast Ambush

December 2000

"**E**vidently he's *always* late," said Aileen Lee, John Doerr's associate. It was almost 8:30 A.M., half an hour after the meeting was supposed to start, and everyone in the locked and guarded ballroom was still waiting for Steve Jobs. The December 8 meeting at the Hyatt Regency near the San Francisco airport had been Doerr's idea. He wanted Dean to brainstorm about Ginger with him and some friends, including Jobs and Jeff Bezos. The three billionaires could spare only a couple of hours, so Doerr's request required a long trip for a short meeting.

Brian Toohey didn't mind. Barely settled in as Ginger's new vice president of regulatory affairs, he was still dazzled by Dean's roster of acquaintances and it was worth some inconvenience to meet these West Coast business icons. Tim Adams and Mike Ferry felt a bit more jaded and exasperated. Traveling to and from San Francisco chopped two days out of a schedule with no fat in it. Tim and Mike also suspected, as did Dean, that Doerr was setting them up for an ambush on his home turf. But all of them also realized that people who invest $38 million sometimes need their hands held, so Tim, Mike, and Brian had each put together a PowerPoint presentation for what Tim called "another dog and pony show."

In addition to Jobs and Bezos, their audience would include Bob Tuttle, Dean's top lieutenant; Michael Schmertzler, representing the $38 million investment of Credit Suisse First Boston; Bill Sahlman, professor of entrepreneurial studies at Harvard Business School and the yenta who had introduced Dean to Doerr and other investors; and Vern Loucks, a minor investor in Ginger as well as a board member. Schmertzler had changed his mind about not coming, probably because of his evergreen suspicions of Doerr.

Brian, keyed up, got to the ballroom early to check the audiovisual equipment. By the time the others arrived, he had filled the screen with a giant photo of Dean, wearing jeans and sitting on an iBOT, smiling widely as he shook President Clinton's hand in the Oval Office.

The smile was missing as Dean pushed a tall hotel luggage carrier into the ballroom. The carrier held a couple of large black duffels, oddly protuberant, and some taped-up cardboard boxes, including an old Apple computer box. Dean instructed the security guard to lock the ballroom doors and not to let anyone enter without permission from someone inside.

When the doors were locked, he opened the duffels and the boxes, removed a couple of chassis and control shafts, and assembled two D1 Gingers using a screwdriver and hex wrenches. He finished in ten minutes, turned one on, and began tearing around the ballroom, looking happier with every revolution. Jeff Bezos arrived. Dean zipped up to him, stopping sharply at his shoe tips. Bezos didn't flinch.

"See how much I trust you?" said Bezos.

"Is that good judgment?" said Dean.

Bezos claimed the other Ginger, and his laugh soon gusted through the ballroom. Doerr entered wearing casual clothes and old sneakers. Dean surrendered his Ginger to him. Everyone was having too much fun to mind Jobs's tardiness.

Dean didn't mind either, for other reasons. He had flown his jet to San Francisco yesterday, carrying the Gingers. A limo hired by Doerr had whisked him and the machines to Jobs's house, where the two of them spent the afternoon. Jobs did most of the talking. Ranting, really, about Ginger's design. So Dean more or less knew what Jobs was going to say today and wasn't in a great hurry to have the Ginger guys hear it.

The others were so intent on Ginger that they didn't notice Jobs walk in. He was dressed even more casually than Dean, in sneakers, a black turtleneck, and Levi's in which a white pocket poked out of a big front hole. There was a hole in his wallet pocket, too. Within a couple of minutes, after some quick introductions, everyone settled around the big square table, Jobs at one corner, flanked by Dean and Doerr.

"Good morning to everyone," said Tim, smiling at the front of the table. "Before we start, we'd like to ask you to hold your questions until after each presentation."

"Yeah, *right!*" snorted Bezos, followed by that honking laugh.

"Otherwise we might as well not be here," said Jobs.

"How long is your presentation?" asked Doerr.

"Each pitch is about ten minutes."

"I can't do that," said Jobs. "I'm not built that way. So if you want me to leave, I will, but I can't just sit here."

Tim studied Jobs for a moment, then turned to the screen and put up a spec sheet about Metro and Pro. "As you can see—" began Tim.

"Let's talk about the bigger question," interrupted Jobs. "Why two machines?"

"We've talked about that," said Tim, "and we think—"

"Because I see a *big problem* here," said Jobs. "I was thinking about it all night. I couldn't sleep after Dean came over." There were notes scribbled on the palm of his hand. He explained his experience with the iMac, how there were four models now but he had launched with just one color to give his designers, salespeople, and the public an absolute focus. He had waited seven months to introduce the other models. Bezos and Doerr nodded as he spoke.

"You're *sure* your market is upscale consumers for transportation?" said Jobs.

"Yes, but we know that's a risk for us," said Tim, "because we could be perceived as a toy or a fad."

If they charged a few thousand dollars for the Metro and it was a hit, said Jobs, they could come out with the Pro later and charge double for industrial and military uses.

Tim's eyebrows shot up approvingly. He looked at Dean, whose face was a mask, so he turned elsewhere. "Mike?" he said, looking at Mike Ferry for a marketing opinion.

"It's a good point," said Mike, giving his usual noncommittal response.

"What does everyone think about the design?" asked Doerr, switching subjects.

"What do *you* think?" said Jobs to Tim. It was a challenge, not a question.

"I think it's coming along," said Tim, "though we expect—"

"I think it *sucks!*" said Jobs.

His vehemence made Tim pause. "Why?" he asked, a bit stiffly. "It just does."

"In what *sense?*" said Tim, getting his feet back under him. "Give me a clue."

"Its shape is not innovative, it's not elegant, it doesn't feel anthropomorphic," said Jobs, ticking off three of his design mantras. "You have this incredibly innovative machine but it looks very *traditional.*" The last word delivered like a stab. Doug Field and Scott Waters would have felt the wound; they admired Apple's design sense. Dean's intuition not to bring Doug had been right. "There are design firms out there that could come up with things we've never thought of," Jobs continued, "things that would make you *shit in your pants.*"

There wasn't much to say to that, so after a pause Tim began again: "Well, let's keep going, because we don't have much time today to—"

"We *do* have time," said Doerr curtly, changing his own ground rules. "We want to get Steve's and Jeff's ideas."

"The problem at this point is lead time in our schedule," said Tim.

Jobs snapped his head from Doerr on one side to Dean on the other, as if he'd been slapped. "That's *backwards,*" he said, his voice rising. "*Screw* the lead times. *You don't have a great product yet!* I know burn rates are important, but you'll only get one shot at this, and if you blow it, it's over." Agitated, he turned to Bezos. "Jeff, what do you think?"

"I think we'd do a disservice to the machine if we didn't give a great design firm a chance," said Bezos in a calm, soft voice, trying to lower the volume. "I think Steve is right—that as he so elegantly put it, they could do things that would make us shit in our pants." Jobs grunted.

After another pause, Tim moved on to the issue of service, determined to move ahead despite the punches coming at him. Within two sentences, Jobs was on him again. Tim put up his next slide, about the new plant, but again Jobs came at him with a flurry of half-insolent questions. Where are you building a plant? *Why* are you building a plant? Why are you manufacturing the machine yourselves?

Partly, explained Tim, because giving our code to someone else would be a great risk. Not a good reason, in Jobs's view, because the code could easily be reverse-engineered. No it couldn't, said Tim. *Could,* said Jobs. He added that Tim should be spending money and management time on other things, especially since there was no way he could convince any world-class manufacturing and procurement people to move to New *Hampshire,* for God's sake, his tone implying that only slow-witted rubes could bear such a place. Dean lifted an eyebrow.

"We have an adequate staff," said Tim defensively, but it sounded as weak as the adjective. Tim had lost control of the meeting. That was probably Doerr's plan all along. Dean sat silently, offering no help or defense as Jobs rampaged through Tim's presentation.

Brian Toohey spoke next, on the regulatory obstacles Ginger would face and how he intended to overcome them. Brian was a big, burly man who knew how to boom his voice, which may explain why he got two minutes into his spiel before Jobs began interrupting. Doerr suggested that instead of going through each slide, everyone should "take a study hall and read the deck" that Brian had handed out, then ask questions. Bezos had already read it, so he started chatting quietly (for him) with Dean.

"Jeff, have you read the *entire* deck?" said Doerr in a schoolmaster's voice.

"Yes, John, I *have,*" said Bezos, amused.

When the study hall ended, Bezos held up Brian's handout. "I think this plan is dead on arrival," he said. "The U.S.A. is too hostile." The "car guys" were going to lobby against Ginger and they were going to win.

"No they're not," said Brian, smiling.

Bezos suggested starting slow, using one city or country as an experimental station. Once Ginger's benefits were clear, the company would have a wedge to pound into U.S. regulations. The perfect place to begin, thought Bezos, was Singapore. "You only have to convince one guy, the philosopher king, and then you have four million people to test it."

Vern Loucks, who had been quietly watching the fireworks up to this point, said, "You mean Goh Chok Tong. He's not a king, he's the prime minister. I can get us in to see him if we want to do that," he added.

Michael Schmertzler hadn't said much. Now he asked when they should instigate a strategic leak to arouse interest in the product.

But Jobs was still shaking his head at Bezos's suggestion. Because of the Internet, he said, slow was no longer possible. People would learn about Ginger in a flash of bits and bytes, and would want one *now*. So a small launch in a foreign place was foolish, because if the machine was unavailable in the United States, the company would blow its chance for $100 million of free publicity in its biggest market. Plus, Singapore was a nest of pirates, and the company would end up spending a fortune fighting them. If the company wanted a slow, controlled launch, better to start on a handful of U.S. college campuses.

"If you show this to Hennessy," Jobs said to Doerr, referring to John L. Hennessy, president of Stanford University and a world-class engineer, "he'll shit in his pants." Evidently Hennessy did that more readily than Jobs did. "And if you offer to give him a hundred of them if he'll run a safety study and a usage study, that's a done deal in ten minutes," continued Jobs. "You do that at ten colleges and maybe at Disney, so people can see them but not buy them."

But he warned that even this sort of slow launch was filled with dangers. If one stupid kid at Stanford hurt himself using a Ginger and then announced online that the machine sucked, the company was sunk, because there was no way to control that or counter it if people couldn't ride one for themselves. With a big fast launch, on the other hand, a few malcontents wouldn't be heard above the general hoopla. "I understand the appeal of a slow burn," he concluded, "but personally I'm a big-bang guy." For the first time that day he smiled. "The risk with a fast burn," he continued, "is that it exposes you to your enemies. You're going to need a *lot* of money to fight thieves."

"We have a few things they can't get," said Dean. "Specialty components with only one source."

"They'll figure out a way around that," said Jobs.

"I've spent nine years looking," said Dean, "and I don't think so."

"I think the emphasis of this conversation is wrong," said Bezos. "You have a product so revolutionary, you'll have no problem selling it. The question is, are people going to be allowed to use it?"

Jobs said he lived seven minutes from a grocery and wasn't sure he would use Ginger to get there. Bezos agreed. Schmertzler wondered if it might be wiser to start with commercial sales. Bezos liked the idea— it was safer and could give the business a solid foundation for growth.

By then it was 10:30. Bezos and Jobs had to leave. As they stood, Dean rose too. He had been almost silent, listening to Jobs like everyone else. Now he thanked Jobs and Bezos for coming. "This is the most energetic discussion we've ever had," he said, "and like all good energetic discussions it leaves you with more questions than answers, and leaves you questioning everything you thought you knew." He paused. "And that's good."

As an inventor, he meant it. But uncertainty was not an investor's favorite state, so Dean wanted to be sure that Doerr and Schmertzler hadn't lost sight of two things through the smoke of Jobs's bombardment. All the talk about small launches at colleges or in foreign cities or among industries worried him. So he reminded them about the serious problems that Ginger could help solve.

"If we really believe that this is a big idea," he said, "then we need to remember that what big ideas do is make new ways to see things. We shouldn't try to put a big idea into a niche." He looked around the table. The first person to speak—surprise!—was Jobs.

"I don't worry about the big idea," he said, "because if enough people see the machine, you won't have to convince them to architect cities around it. People are smart and it'll happen. That's the story of the PC. Nobody had any idea how they would be used, and look what happened."

"I think we all agree with the big idea, Dean," said Bezos. "That's why we're here, not to make $3 billion in the golf cart market."

Very nice, but neither Jobs nor Bezos had a dime invested in Ginger, and Dean hadn't heard from anyone who did. Maybe it was time to throw a bone. "But you can still make money on the niche markets if you want to," said Dean, patting Schmertzler on the shoulder.

"Dean, we've all made plenty of money," said Doerr, sounding offended. "We're not here for the money. Please retract your comment."

How rich. Here was a $38-million investor acting as if money didn't matter, like a gambler lighting a cigar with a $1,000 bill. Dean smiled and apologized for suggesting that his investors might prefer to make money instead of changing the world. Schmertzler shifted his eyes from Doerr to Dean, wearing a look that mixed disbelief with consternation. Changing the world would be nice, said the look, but first let's recover our investment and compound it.

Jobs left but Bezos decided to stay for Mike Ferry's marketing report. Though Bezos said he wouldn't use Ginger at work or to run errands because he liked to walk, he loved the machine and *would* want one to show friends. He didn't like the marketing plan's emphasis on college campuses, where Ginger would be competing with bikes, which were cheaper, faster, provided exercise, and didn't need charging. Better to launch in a big city, he said, where people wanted cars but couldn't afford them—a place like Singapore.

Aileen Lee, Doerr's young associate, spoke for the first time. Could Mike Ferry state Ginger's "value proposition"? Trendy M.B.A. jargon for "gimme the headline." Lee had made a list of significant new products and assigned each a value proposition. Next to cell phone, for instance, she put "allows wireless telephone communication." Next to fax machine she put "allows visual transmission of documents over phone lines." And so on for the PC, the PDA, and other products. What, she asked, would Mike put next to Ginger?

Mike seemed flummoxed and began rambling about the different markets for the machine.

"*Answer* her *question,*" said Doerr sharply. "What would you put next to *us*?" He paused infinitesimally, then bored in again. "You're the marketing guy. You should know." Another tense pause. "*I'll* answer if you can't."

"I think it's a really good question," said Mike blandly. "And—"

Doerr snorted in disgust and shook his head, then looked at Dean to see if he realized that his marketing director was a dinosaur.

Bezos and Doerr had to leave, but Doerr still seemed keyed up. He had drawn several conclusions from the meeting, he said. First, they needed a value proposition. Second, until they got the product right, the launch was indefinitely on hold. Bezos asked Dean how he felt about that.

"I think the design can be fixed with relatively small impact on the launch date," Dean said carefully.

"We get to do something like this so rarely," said Doerr, sounding righteous again. "We shouldn't launch until Steve Jobs wets his pants."

"Until he *shits* his pants," said Bezos.

"It's going to be a messy meeting," said Loucks.

"Especially if we all agree with him," said Dean.

Dean didn't doubt Doerr's noble motives, but he knew that if they missed the launch date, Ginger would need more money, because the burn rate at that point would be millions of dollars per month. Money might not matter to Doerr at the moment, but it always mattered to Dean. If Ginger needed more funding, Kleiner Perkins and Credit Suisse First Boston would offer it eagerly—for a bigger share of his company. That wasn't going to happen, whether or not Steve Jobs thought the design sucked.

Doerr, Bezos, and Schmertzler left. The group took a break. Mike Ferry looked agitated. Aileen Lee's question about a value proposition felt like a cheap shot to make her look good. Tim Adams seemed relaxed and was smiling. He found the idea of a value proposition ridiculous at this point; they hadn't even done any significant consumer testing. It was equally ridiculous to suggest delaying the launch until Steve Jobs befouled himself. The meeting had confirmed the Ginger team's suspicions that the whole thing had been a setup. They hadn't been invited all the way to California so that Jobs and Doerr could pat them on the back. They had expected to get creamed no matter what they said, and they had been right. The meeting resembled the carnival game where the goal is to pound down every head that pops up.

Doerr had arranged a parade of other tourists. Ray Lane showed up first. A short, trim, practical-looking man, he had helped Larry Ellison build Oracle but had recently joined Kleiner Perkins as a partner. Randy Komisar bustled in wearing faded jeans, black cowboy boots, and a black leather jacket loaded with zippers. Komisar called himself "a virtual CEO" and had advised several Silicon Valley start-ups. Bill Campbell, a wiry guy with a raspy voice, was the CEO and chairman of Intuit. They all signed confidentiality agreements and took Ginger for a spin.

Dean launched his spiel about the Big Idea and cars and pollution, then showed the homemade Ginger video, narrating it as he had dozens of times. He was a superb salesman partly because he never tired of making the pitch.

Komisar loved the technology. He predicted that Ginger would be ubiquitous someday, but that changing people's attitudes and getting on sidewalks would take a decade. He counseled patience. If Dean tried a fast ramp-up by "spending hard against the adoption curve," he would waste a lot of money. Instead, go slow and build a track record at places such as Disney or national parks or police departments.

Campbell's succinct advice: "Get a Japanese partner *tomorrow*."

Bob Tuttle asked their opinions of the machine's design and aesthetics. Not bad, said Komisar. Acceptable, said Campbell. Sahlman told them that Jobs thought it sucked.

"That's just Steve," said Komisar.

"I've seen him just *kill* his designers," said Campbell, who sat on Apple's board. He asked how Jobs had liked riding Ginger.

"The first time he rode it," said Dean, "after thirty seconds he said, 'This is the most amazing piece of technology since the PC.'"

Bob Tuttle, Bill Sahlman, and I flew back to Manchester in Dean's jet. Cruising at 41,000 feet, Dean did some post-meeting analysis of all the contradictory advice. He thought Jobs was wrong about the ease of reverse-engineering Ginger, wrong about building a plant, wrong about New Hampshire. He thought the group had been unfair to Tim and Mike, and dismissed Doerr's comment about delaying the launch as "just being dramatic."

We were flying into the sunset. Dean started talking about the troposphere and the curvature of the earth and ions and bouncing radio waves. We're flying at 460 miles per hour, he said, eight miles above the earth, in 12,000 pounds of aluminum. The temperature outside is minus 144 degrees. "Think about it," he said, savoring the technology. "And we're listening to the stereo." Yes, a CD of Linda Ronstadt singing standards that repeated four times between San Francisco and Omaha.

But he was mostly quiet on the long flight home. There were so many things to think through, including damage control among his staff and his investors. His biggest worry from the meeting wasn't Ginger's design or regulatory obstacles or distribution channels or where to launch. It was his management team. For months, Doerr had been calling Tim and Mike mediocre. Jobs and Doerr had been unfair today, but Tim and Mike had seemed outmatched, stodgy, *traditional*. Only Brian Toohey had done well.

Within a few days, Dean's fears were confirmed in phone conversations with several people from the meeting. Doerr, Schmertzler, Sahlman—all of them asked the same thing: Why was Dean letting these guys run his company? Doerr attacked Mike and insisted that Ginger needed a value proposition. When the PC appeared, retorted Dean, nobody knew its value proposition. Would it be used for adding numbers? Interoffice communication? Writing business letters? Tracking inventory and sales? It was premature to ask for Ginger's value proposition. But he did know that every major city had at least two big problems: pollution and congestion, caused mostly by cars. And he did know that Ginger was the ultimate pedestrian and would have a huge impact on transportation, on the way people lived in cities, and on the worldwide economy.

He was delivering the spiel again, and Doerr asked him to stop. Why could Dean articulate those things, Doerr asked, but his director of marketing, who had been there for a year, couldn't think of *anything* compelling to say?

For once in his life, neither could Dean.

chapter sixteen

Fallout

Late December 2000–Early January 2001

A few days after the San Francisco meeting, Dean drove to the airport to pick up his old friend Raymond Price. A distinguished, modest man who had been both a speechwriter and a ghostwriter for Richard Nixon, Price was now president of the Economic Club of New York. He had brought a friend: Howard Burson of Burson-Marsteller, one of the world's largest public relations firms. Burson was a potbellied elf with wild gray eyebrows, oversize glasses, and a black suit jacket with thick red stripes. He had needed a haircut for some time. In his seventies and semiretired, he still held power within the company.

Making conversation on the drive back to DEKA, Dean asked whether Burson-Marsteller generally worked "for the good guys or the bad guys."

"Both," said Burson. "Which are you?"

"We've figured out a way to put nicotine and alcohol into chocolate milk," said Dean, "and we need your help."

At DEKA, Mike Ferry joined the group in Dean's office. Dean wound up and gave Burson the full pitch. Near the end, Dean brought up his concerns about one of the agency's clients—Ford. He worried that Ford might require Burson-Marsteller to drop Acros or even reveal Acros's strategy. Ford wasn't a big client, said Burson. They only spent about $1 million a year. But a couple of months ago, in October 2000, his agency had joined the WPP Group, a conglomerate of eighty companies that included some of the world's largest advertising firms—Ogilvy & Mather, J. Walter Thompson, Young & Rubicam—as well as the world's other huge PR firm, Hill and Knowlton. And Ford did spend a quarter of a billion dollars a year with WPP.

"A quarter of a *billion*?" said Dean. He gulped his coffee, which had grown cold during his pitch, then looked at Mike Ferry. "That's even more than *you* want to spend."

Ford prohibited WPP from working for any other automotive company, but Burson didn't think Ginger fit that category or would even be on Ford's radar for six months to a year. Dean asked what Burson would do if Ford called the agency a year after Ginger's launch and told them to stop helping Acros.

"That's easy," said Burson. "We'd drop you."

Clients sometimes did make such demands. Burson pointed out that potential clashes were inevitable if Acros wanted to use a big agency, and that Dean hadn't mentioned other Burson clients who might find Ginger irritating—Shell, Chevron, and British Petroleum.

Dean thought Ford might catch on faster than Burson imagined. J. T. Battenberg from Delphi had understood Ginger's implications immediately. If Battenberg got it, so would Ford and the other car companies, especially since they were planning to expand into developing countries.

"And I'm sure that because my brain and my mouth work together," said Dean, "I'm not going to be able to help saying, 'Why do you put your 150-pound ass in a 4,000-pound vehicle to move around?' And I won't be able to help saying, 'Do you know there will soon be 4 *billion* people living in cities, and there's no reason that anyone who only needs to move ten or twelve blocks should own a car?' I won't be able to help saying that because it's the truth, and it's what makes our product compelling. The auto companies are going to hear that and understand what it means."

Burson thought Dean was fretting too much. Ginger wasn't going to replace the car. But Burson agreed that if Ford squeezed his agency, "You'd lose."

Dean looked thoughtful. If you were me, he asked, what would you ask from Burson-Marsteller? Acceptable notice, said Burson. Maybe a year, with the right to take the account people. That sounded tolerable to Dean, if Burson would guarantee it in explicit language.

"By the way," said Dean, "4 percent of the world's population owns cars. The other 96 percent are *my* customers."

In the hierarchy of his paranoia, Dean ranked Ford third. Honda came first, because it made small vehicles, followed by Sony, which made electronics. But he wondered who else on Burson's roster might be problematic for Ginger?

Sony was a client, said Burson, but he didn't think they posed the same kind of problem. He began to tick off the agency's biggest clients. When he mentioned the U.S. Postal Service, Dean jumped in with a question: Could you get us into the post office? He had been salivating about the federal agency's huge workforce since the early days of Ginger, but hadn't found a way in. Maybe fate had delivered the key to the postal kingdom.

"I could pick up the phone and have you in Bill Henderson's office next week," said Burson.

William J. Henderson was the U.S. postmaster general. But he was leaving office in May 2001, said Burson, so Dean would need to move fast. Burson added that Henderson supervised the country's largest payroll—eight hundred thousand employees. Dean looked like a bird dog quivering on point. He wanted to see Henderson right away. And if Henderson was leaving soon, that might be even better, said Dean, because he would know everybody and everything about the Postal Service, and could work for Acros as a consultant. Dean wanted to set it up immediately. Burson offered to go to Washington with him.

Later, Burson took a tentative ride on Ginger.

Ginger excited him. He promised to get Acros the agency's best available team. Brian Toohey, Ginger's director of regulatory affairs, was gleeful at the prospect of inserting Ginger into the Postal Service. He called it a clear winner in every way. Dean agreed. The police weren't going to tell mail carriers to get off the sidewalk.

The fallout from the San Francisco meeting kept dropping like fine ash. Jobs's excoriations had spurred Dean to call an acquaintance, Bill Stumpf, the designer of the Aeron chair, to meet about Ginger's design. He also hoped to show Ginger to David Kelley, founder of IDEO, a Silicon Valley design firm. Michael Schmertzler, John Doerr, and Bill Sahlman all continued to press Dean to cut loose Tim Adams and Mike Ferry.

Dean was torn. He valued loyalty and hated to fire people, but his allegiance to Tim and Mike was weakening. "I think they're clearly out of their depth and don't belong working for a start-up," he said. "But if this grows as fast as we expect, we'll be so busy with suppliers and distributors that we will grow into their skill set."

Tim didn't realize how far his stock had dropped with Dean and the investors. When I mentioned to him that he must be feeling bloodied from the meeting in San Francisco, he looked puzzled. He hadn't felt that way at all. Yes, Jobs had been aggressive and noisy, but Tim had been through similar encounters with big egos at conference tables. He didn't feel beaten up. When Jobs started windmilling, he had simply backed off. No real damage done. After all, he and Dean and Bob Tuttle had talked beforehand about what might happen in San Francisco. They knew it was an ambush, and Dean had talked tough about not letting Doerr and his gang dictate anything.

Tim *had* been a little surprised when Dean just sat there while Jobs and Doerr hurled grenades. Since Dean had spent the previous day with Jobs, the troubling thought had crossed Tim's mind that Dean knew what was coming and didn't warn him. But he assumed that he and Dean were still on the same team and had been following the same strategy: Let the West Coasters squawk, then go home and get back to work.

Tim always took the rational view and expected rationality from others, so for him the meeting in San Francisco hadn't been personal. But for Dean everything was personal, because everything about Ginger was an extension of who he was and what he believed. Tim badly underestimated that. More important, he underestimated how far Dean would go to keep his investors pleased and confident that he took their interests seriously.

Doug Field also had heard some excerpts from the San Francisco meeting. He asked for my take. It was probably better that you weren't there, I said. Surprised, he asked why. Because, I said, certain remarks might have upset you. "You mean like Steve Jobs saying that I suck as a design manager?" said Doug, smiling. "I don't take that seriously."

Still, it must have stung. Doug admired Jobs's fierce attention to design, so his opinion did carry weight, even though it was directed at D1, a rev receding into the past. Doug admitted as much by deciding not to relay Jobs's remarks to his team. Too demoralizing, he said.

A week after San Francisco, Tim Adams called another town hall meeting to inform the team about the company's financial condition and projections. Just before it began, Dean showed up and huddled with Tim and Doug, looking unhappy. Dean had forgotten about the meeting, and when he happened to stop by Ginger and heard about it, he had rushed downstairs. When he learned that the topic was money, he wasn't pleased. The nerve of Tim, acting like Ginger's CEO.

Using an overhead projector, financial director Scott Frock told the team that the list price on the Metro and Pro had been raised several hundred dollars because the cost of materials was still higher than expected. Frock projected sales of $350 million in 2002, the launch year, with a substantial profit. These assumptions were based on an estimated "cost of goods" that the team hadn't yet achieved. The burn rate now was $4 million per quarter. By the launch date in early 2002, that figure would at least double.

An engineer asked whether the company would do another round of financing if these estimates were incorrect and the company needed cash at the end of 2002, the first year of production. Frock said he understood that Dean would go to the banks for a loan rather than give the investors a bigger piece of the company. Dean had been lying on the floor listening. He sat up.

Doug asked when the investors wanted their money back. Dean stood up. Without answering the question, he said the investors wanted a ten-times return—1,000 percent. This number seemed to open his spigot. "We will *cherish* those days when we were a private company," he said. "If I had enough resources, I'd fund this myself, as I have every business since college."

He didn't understand that that was exactly what many members of the team didn't want. If Dean could fund Ginger or buy out the investors, the company would never go public, the employees' stock options would be worthless, and Dean would retain complete control of their financial and professional futures.

Dean continued: Acros had received nearly $90 million without giving up much control, "which is almost unprecedented." But if the launch slipped a quarter or two and Ginger needed another $15 or $20 million, the investors would be at his elbow offering help—for a larger chunk of the company, which meant less for everyone sitting here.

"They're not bad people," said Dean. "They'll look me in the eye and say, 'You didn't do what you said you would.' If we run out of money, it will be ugly. That's why I may seem grumpy, because it scares the shit out of me, and we've already had a couple of big 'oops.' But I still think this is the most exciting product I've ever seen."

He started selling them on Ginger. It would be world-shaking. It would help solve significant problems. They were heroes for building it. It could be as important as the phone or the car or the PC. But they had to control the schedule and expenses. He told them that the next time they wanted to hire more help, to remember that the cost wasn't just another $50,000 salary. Stock was priced on a multiple of earnings, so another $50,000 expense actually subtracted fifty or a hundred times that amount from the company's valuation. "That should be *viscerally* part of your thinking."

He paused to let that sink in. He disliked talking about these things, he continued, because they distracted people from the real job—launching Ginger. But if the team slipped, the investors would pounce. "It all depends on the people in this room," he said, slowly looking from one edge of the group to the other. He rarely quoted anyone, but now he offered a favorite saying from General Douglas MacArthur: In war as in every other endeavor, failure can be summed up in two words: too late.

So much for John Doerr's decree that Ginger's launch was indefinitely postponed.

"You're a little group of people trying to do something that no rational people would try," said Dean, leaning against the wall. "Unfortunately, I don't get to see you very often and I'm embarrassed to say that I don't even know some of you. This is my favorite project and I don't get to do much of the fun stuff on it."

It was his usual evangelical mixture of complaint, praise, warning.

The room had gotten very quiet. Tim broke the tension by inviting everyone to a local tavern for beer.

At the bar, reactions varied to Dean's honey-and-brimstone homily. A new guy who had never heard Dean speak said, "It scared me to death." A more cynical veteran asked if I intended to do a follow-up book of Dean's collected sermons. The speech had made Ron Reich's engines race (most things did). He was determined to meet the deadline, not just because he could benefit financially, but for the team and for Doug. "I would bet money that anyone on this team would do *anything* Doug asked them to," he said.

A beer or two later, Dean and I walked back to his office. He went through his mail. Christmas cards, including one from Paul Pressler, president of Walt Disney Attractions. A magazine—*Harvard Business Review*—straight into the wastebasket. And a packet of information about a new jet, the Raytheon Premier I. Bigger and faster than his Citation, it could fly nonstop to the West Coast. He had sent a down payment and expected delivery by early 2002. Just in time for Ginger's launch.

Chip MacDonald and three other members of the manufacturing team needed to travel to Detroit to meet with the company that would be building the production line. As a budget measure, their boss, Don Manvel, suggested that instead of flying, they rent a van and drive. Sure, they joked, we'll rent a Winnebago so we don't need hotel rooms, and we'll take a hibachi so we can cook all our meals. But Don was serious. He had become infected with Dean's neuroses about expenses. The others couldn't believe it. No way they were driving to Detroit from New Hampshire, especially in the middle of winter. The usual round-trip airfare from Manchester was $1,000. Chip found a fare for $350 that took them to Chicago and eventually backtracked to Detroit. It took a lot longer, but it was a bargain.

Doug was trying to shave costs, too, ones that mattered. Near the end of December, he, Ron Reich, and Mike Martin listened to five people from Axicon report about their sound tests on seven transmissions. Axicon needed a decision that day in order to start making one hundred fifty transmissions for D3. The choices came down to transmissions with helical gears or with spur gears. Helicals were smoother and quieter, but also $8 more per machine. Doug loved them. Ginger's tone was important to him.

Ron started working his way through the pros and cons of the seven choices, but didn't get far. "Helicals are off the table," Doug blurted. "I'd love to have them, but we can't do it."

The Axicon guys looked disappointed. They were engineers too, and were proudest of their helical gears. That left spurs, which came in two versions, hammerhead and conventional. The Axicon guys passed around their new patent-pending hammerhead. Doug oohed and aahed, fondling it.

"The hammerhead is $2 more per machine than the conventional," said Ron.

"$2!" said Doug, his hands suddenly still.

"No, it's 86 cents times two," said an Axicon guy.

"Then it's easy," said Doug, putting the transmission back onto the table. Ginger would launch with the conventional gears. Time to destroy, not create. The Axicon guys looked even more disappointed. After Doug left, they said they couldn't believe he wouldn't spend an extra $1.72 for better gears. Well, said Ron, they hadn't heard yesterday's town-hall speech about cost containment.

Costs and budgets were on my mind as well. After working on spec at Ginger for nearly a year and a half, I needed money. Dean was sympathetic. The week before Christmas, with Dean's approval, my agent sent a handful of editors a book proposal that amounted to an elaborate tease, offering enough details to tantalize them without revealing the nature of Ginger.

It snowed on the night of DEKA's Christmas party. Dean met people at the door, worried that they might have had difficulty getting up his steep driveway. As always, the food was delicious and plentiful—shrimp, crab-stuffed mushrooms, roast beef, salad bar, dessert table.

This year's skit, *It's a Wonderful Company*, was based on the classic Christmas movie *It's a Wonderful Life*, starring Jimmy Stewart. In the DEKA production, Bob Tuttle took the voice of God. The story's setup was a slide of Dean receiving the National Medal of Technology, where the necessity of wearing a tuxedo sparked a dreadful vision of a future plagued by memos, organizational charts, and more tuxedos.

What would be the fate of people at DEKA, asked the skit, if Dean disappeared? The skit milked this prospective calamity for laughs, but the stark subtext was everyone's absolute dependency. No Dean, no DEKA.

Benge Ambrogi presented the engineers' Christmas gift to Dean—a clock with elliptical gears. As far as they knew, it was the first such device ever made. Benge explained how it worked: The leisure time between 6:00 P.M. and 6:00 A.M. sped by in two hours, while the work time between 8:00 A.M. and 4:00 P.M. dragged on for nearly twenty hours. But since this design also stretched lunch into almost three hours, which Dean would find intolerable, the engineers had added another gear that crushed 11:30 to 12:30 into thirty minutes. Benge demonstrated by winding the clock's hands, which barely moved during work hours but whipped around the clock face during off-hours. Dean laughed hard.

A nine-piece band from Boston started playing swing music. The Christmas bonuses were twice the size of last year's. Dean had paid off his mortgage again. What a difference a year can make.

Kevin Webber was hiding behind Bill Arling, who moved another 10 feet away. Kevin ducked behind him again. "It can really explode," said Kevin, "so I'd stand back."

They looked like kids in chemistry lab, but the location was Mike Martin's torture chamber. A control shaft/handlebar was hanging up-side down against a wall. From one of the handlebar's grips hung a small shelf filled with weights. An engineer named Pat Kelly, wearing safety glasses and looking a little tense, was kneeling next to this arrangement with another 10-pound weight in his hand. He inhaled and slipped it onto the shelf—that made 140 pounds. Silence. Exhale.

"So the newest guy gets to put the weight on?"

"Well, he designed it," said Bill.

150 pounds. Exhale. 160 pounds. Exhale. 170 pounds. Pat eased a 5-pounder onto the shelf. 175 pounds.

Crack! Everybody jumped. The handle had sheared off. The engineers huddled to examine the fiberglass-reinforced plastic. Not bad for a first try on an inferior mold. Their goal was 200 pounds, the most extreme probable load, unless someone was intent on breaking the bar.

Doug bustled by, heard the number 175, and paused to shake Pat's hand. "I'll sleep better this weekend," he said.

Pat transferred the hanging shelf to the other handle and started adding weights.

More fallout. Jeff Bezos called Dean for a long talk about San Francisco. Bezos laughed about Jobs's outrageousness and asked again if he could be involved with Ginger. Dean intrigued him, he said, because he didn't seem to be motivated by money. He worked on projects that could help people—medical devices, FIRST, Ginger. And the Stirling engine—the obvious market was the United States, said Bezos, but Dean envisioned it providing power and clean water in developing countries.

I hate to disillusion you, said Dean, but I'm not the Mother Teresa of technology. I don't think money is a four-letter word and I like to make it, and I've done OK.

Of course, said Bezos. After all, he had seen Dean's little house on the hill. But his point was that money wasn't Dean's primary motivation. Like John Doerr, Bezos wanted to be involved because of Dean's broad vision, but Dean needed people who understood that vision and how to implement it through a business. Tim Adams and Mike Ferry weren't those people, said Bezos. Middle managers, maybe, but not leaders of a new international business. Dean offered his usual defense, that the company would grow into Tim's and Mike's skill sets.

Your problem isn't John Doerr, who will do whatever you want, said Bezos. Your problem is *you*. You know John's right and you don't want to face it.

As Dean told me all this, he repeatedly ran a hand back through his hair, his usual tic when troubled. He interspersed the story with parenthetical remarks about how perceptive and self-effacing Bezos was, how much he seemed to care about Dean and Ginger. When Dean was talking his way through a problem, you could watch his mental gears clicking. Right now those gears were moving him toward this: If Bezos was right, things at Ginger were wrong.

Michael Schmertzler had been saying the same things. Bill Sahlman, too. "And he's such a low-key, easygoing guy," said Dean of Sahlman. "For him to say it and be so certain of it . . ." He shoved a hand through his hair.

Early the next morning, Doug came into the cafeteria for coffee. He still claimed to be too busy to dwell on the vague criticism from San Francisco. He was freezing parts and moving ahead.

"I'm getting pretty belligerent about it," he said. "What really irritates me," he continued, "is all the churning, the *endless* churning, about what we should do about what Jobs said. I'm losing patience with that. He didn't say anything specifically helpful and he's not even an investor."

Doug rarely unloaded, and it took a lot to tip him. He was still nettled that Tim had asked him yesterday about doing ultraviolet tests on the dashboard. As if he had forgotten about UV testing. As if he didn't already have a long list of tasks that were much more important.

"There are things I can control and things I can't," he said. "Right now I can't get the phones working. I can't get furniture." The move into the new space had been vexed by such things, adding to the team's stress level. The engineers had run makeshift phone wiring out the windows and back inside. "I wish someone would fix *those* things." He meant Tim.

I offered a theory: The churning occurred because the only opinion that really mattered was Dean's, so people spent a lot of time trying to figure out what was in his head and how he might react to various things.

"Then I should be dealing directly with Dean and not have these interfering filters in my way," said Doug.

"But since Dean seems to pay attention to Jobs, so does everybody else."

"The fact is," said Doug, "Jobs has nothing to do with this organization."

"But he does," I said. "Look how much reaction he's caused."

Doug was quiet for a moment. "You're right," he said.

Dean called me a few days later, on the last Friday in December. He sounded tense and miserable. He had just gotten off the phone with Steve Jobs. Jobs had attacked Tim Adams and Mike Ferry for fifteen minutes. Dean quoted him: They're buttheads. That marketing guy should be selling Kleenex at a discount store in Idaho. And that CEO— where did you find an old-line butthead like *him*? Jobs said he had never been so excited about a new technology, but he agreed with John Doerr—Ginger needed a new management team.

Dean sounded agonized. But Jobs wasn't the reason he had called. He and some of the Ginger guys were going to Delphi–Kokomo the day after New Year's to decide about the gyros, the latest cause of friction between him and the Ginger group.

A few days earlier, in Tim Adams's office, Doug had asked what Tim wanted to do about the gyro dilemma. Don Manvel, director of manufacturing, tensely awaited his answer. BAE Systems, the British company making the gyros for Ginger, had told Don that if Acros wanted the company to gear up for mass production, it was time to put up some money. BAE wanted a financial commitment, including a letter of credit for millions of dollars. That reversed Dean's preferred direction of cash flow, so the request affronted him. He told Don to put BAE off and pressed Delphi to come up with a cheap gyro.

After thinking about Doug's question for a moment, Tim said he wanted to stick with BAE for launch. He didn't believe that Delphi could deliver the gyros on time, in quantity. Betting on Delphi could wreck the launch, whereas BAE was a sure thing. "It doesn't matter if we reduce cost if we don't have inventory," said Tim. Doug pumped his fist in agreement. Don looked relieved.

Dean had shrugged off their unanimous opinion. He continued to sweet-talk Delphi and to balk at committing to BAE. That left Don in an impossible position. He had toiled to get the best deal and the lowest price from BAE, and the company had spent a pot of money to develop a special gyro component for Ginger. Now Dean seemed willing to leave BAE at the altar and let Don deliver his farewell.

On the other hand, maybe the bride should have seen it coming. Six months earlier, Dean had jilted another gyro manufacturer, Systron Donner, when BAE offered a better price for the gyros used in the iBOT. At the time, an engineer described Dean's way of dealing with suppliers this way: "Dean is very good at dating two people at once. But I wouldn't want to be there when Systron finds the second pair of underpants in the bed."

Now BAE wanted the gyro contract for Ginger as well, and had asked Dean for a financial commitment before they spent more millions. What impertinence. Nor did Dean like the idea of having a single-source supplier for Ginger's most important component. Tim and Don didn't think BAE would "screw" them. "But of *course* they will," Dean had said. "It's *business*. They wouldn't screw GM, because that's a huge company, but of *course* they'll screw us if they have us by the short hairs. It doesn't mean they're bad guys, it's just the way things work." He sounded cool and rational, like Tim Adams. But the truth was that if Dean felt that someone was trying to screw him, it wasn't just business, it was personal, and he never got over it.

Dean had requested the meeting at Delphi on January 2, 2001, to clarify the gyro problem. On New Year's night, lying on his library couch and sipping a beer, he gave me some background. For years gyros had been an expensive specialty product used mostly by the aerospace and defense industries, but in the late nineties, many companies started developing lower-cost gyros. For technical and design reasons, Ginger needed a component containing multiple gyros (the House of Gyros), plus accelerometers and an inclinometer. BAE wanted to charge Dean 50 percent more per unit than the team had projected.

Yet the Ginger guys weren't seriously looking at other manufacturers, complained Dean. (In fact, the team had looked at many gyros, but by "seriously" Dean meant that they hadn't wheedled a better price from anybody.) Meanwhile, DEKA and Acros had started buying more electronics from Delphi, which had a gyro in development. Dean had asked the company's CEO, J. T. Battenberg, if Delphi could make a gyro unit for less than half the cost of BAE's. Battenberg eagerly agreed to try. "I know it's the eleventh hour," said Dean on New Year's night, "but it could save a lot of money."

But the Ginger guys wouldn't listen. "They circle the wagons," grumbled Dean. "They think, 'This lunatic is bothering us again.'" Delphi had put a team of fifteen on the gyro project, an indication of Battenberg's enthusiasm, and Dean wanted the Ginger guys to work with them. But they begged off, citing Dean's request not to slip the schedule. "It's malicious obedience," said Dean. He had turned the issue over to a couple of DEKA engineers, who were testing various gyros and kissing frogs. "Those guys across the hall at Ginger don't have time to get warts on their faces," said Dean. "They're running a religious cult over there, not an engineering firm. They're more concerned about the process than the result. It drives me *crazy*."

"Over there." "The guys across the hall." As if Ginger was a troublesome colony demanding independence.

The Delphi team wasn't sure it could meet Ginger's deadlines, but the company had dropped its price. BAE, by contrast, would not drop its price and also wanted a guaranteed purchase of a hundred thousand units in the first year and at least 50 percent of all sales in the second year.

"I jumped out of my chair," said Dean. And there was also that multiple-million dollar letter of credit. Tomorrow's meeting at Delphi was costing him $100,000 because of a penalty clause in the BAE contract that kicked in when he hadn't signed the purchase order by the end of the year. He felt hosed.

So he had told the Ginger guys that tomorrow's meeting at Delphi offered four possibilities: (1) You discover that Delphi can't do it; (2) Delphi admits they can't do it; (3) Delphi can do it and you get excited; (4) Delphi can do it but you refuse to believe it. If the last case proved true, he would force Delphi on them.

I asked what he had decided about the investors' criticisms of Ginger's management team. Dismay crossed his face. "I own most of the company, but it's their money, so I like to do what they want," he said. "If I know in my heart that they're wrong, I try to convince them, and I won't do it." He paused. "But on this issue, I'm not sure they're wrong."

He was silent for a few moments. "I felt sick when I got off the phone with Jobs. On the other hand, I also have rational people like Jeff Bezos saying they aren't the guys to introduce a new product and a new industry. Is he right?" He looked tired. "I don't know."

The next morning at 7:20 we lifted off with Dean at the controls. Tim Adams, John Morrell, Phil LeMay, and two DEKA engineers were aboard. Don Manvel and Doug Field were meeting us there.

The Kokomo, Indiana, airport had one runway, black against the flat, snow-covered fields around it. Farms spread out in every direction, cut by two-lane roads with sparse traffic. The lowering gray skies promised more snow. En route to Delphi we passed an Amish man driving a horse and buggy, a study in black. The day was bitter cold and I wondered how he stayed warm in there since he didn't believe in machines. "Hmmm," said Phil, immediately snagged by the technical challenge. "He could run plastic pipe with water in it around the horse and capture that heat, and the wheels could drive the water through the pipe."

At Delphi, two dozen people crowded into a conference room. Dean began by saying that he knew the Delphi gyro was a long shot, but he couldn't resist exploring the possibility, "being the entrepreneur in heat that I am." But if Delphi couldn't absolutely guarantee what Ginger needed, no go. "Because if we can't get the gyros, we have no business."

Dean and the Ginger engineers began grilling the other side of the table. Often, the answer that came back was, "We don't know." It became clear that the Delphites didn't have an acceptable gyro ready. They expected to complete a final test version just before Ginger's scheduled launch. That would be a huge risk. If the test gyro contained flaws, it could abort the launch or cause it to crash. Delphi's time line also looked extremely condensed, with no room for hiccups.

Doug was biting his nails. Tim Adams began pacing around the table, his habit while thinking through an issue. He asked about validation testing and lead times for tooling. The answers didn't please him. "I don't see how we can do it," he said. He kept pacing and probing, his head down, his eyes unblinking as he shot quick questions at the Delphites. He got them to admit that a normal time line for freezing and validation of the gyro would push full production to late 2002, long after Ginger's projected launch. Tim exposed flaws that seemed to demolish the possibility of using Delphi. He had taken over the meeting, and his performance demonstrated the knowledge he could bring to Ginger if given the chance. The Delphi team asked for forty-eight hours to make a last-ditch proposal and for a few minutes now to talk.

The Ginger team settled into a lounge area. Dean asked for people's opinions. John Morrell thought that using Delphi would delay the launch six months. Doug agreed. Phil LeMay didn't think Delphi's gyro was worth the risk. Don Manvel didn't believe Delphi could deliver the volume. Tim's view was clear. It was unanimous, with one exception.

"But if BAE ends up three months behind," said Dean, "then Delphi is only three months behind that."

Tim grabbed a magazine off the coffee table—*Redbook,* of all things—and began snapping through the pages, trying to contain himself. Dean suggested that BAE wasn't without risk. The company was depending on a gigantic corporation, Sumitomo, to build a gyro plant in India that didn't yet exist. Then BAE would be depending on chips that would come from someplace else, and on a board-stuffer to assemble the whole thing. BAE's plan could fall apart at several spots. John Morrell pointed out that the Delphi engineers had dodged many technical questions. Dean agreed; that bothered him, too.

Tim put down the magazine and in a low, heated voice reminded Dean that the Delphites had to be cornered about the schedule and hadn't shown anything that merited confidence. Dean nodded but added that Battenberg wanted this business. To get it, he would put on enough extra people to make Kokomo glow at night. Delphi was one of the best high-volume producers in the world, Kokomo was the world's biggest chip plant, and it was only an hour and a half from Manchester, as opposed to India. The difference between Delphi's price and BAE's could cost Acros tens of millions of dollars over three years. He paused to let them absorb those facts.

Did they really think Ginger would launch in January 2002, he asked, even if BAE's plan went smoothly? He studied each of them in turn. The silence became uncomfortable. Doug finally said, "I can't say." Neither could John. Tim still considered January 2002 feasible. Dean said he was asking because if some other glitch caused a three- or four-month slip, they were only a couple of months away from Delphi's much cheaper gyro.

He was an effective devil's advocate who didn't insist or browbeat. "If I knew the right answer," he said, "I'd tell you." In any case, in forty-eight hours they would know if Delphi had come up with a convincing argument or not. Everyone nodded.

Over dinner in Manchester that evening, Dean outlined other recent developments. One of the millionaires who had seen Ginger last summer and wanted to invest $5 million was now the secretary of defense: Donald Rumsfeld. After looking at Ginger, Rumsfeld had given Dean a list of things to do and had called two weeks later to see if his orders had been carried out. "He scared me," said Dean. He shied away from including Rumsfeld as an investor, but as secretary of defense he could be a wedge into the huge military market.

In fact, Dean knew almost every member of George W. Bush's new cabinet, including another man who might be helpful to Ginger. A few weeks earlier, Norman Mineta, then Bill Clinton's secretary of commerce, had draped the National Medal of Technology around Dean's neck. Now he had become George Bush's new secretary of transportation.

The day after the meeting in Kokomo, Dean sat down in DEKA's conference room with two top executives from Danaher Motion, the parent company of Pacific Scientific, which made Ginger's brushless electric servomotors. These were smaller than a soup can but capable of nearly 3 horsepower, the highest density motors in the world. The iBOT used four of them. Pacific Scientific had worked intensely with engineers at both projects to develop the motors, which had gradually dropped in price from the low four digits for the early prototypes to an estimate in the mid-double digits once mass production began.

That was the reason for this visit. To meet the expected demand from Ginger, Danaher Motion needed to build an assembly plant costing $14 million. But Ginger was new and untested. What if it flopped? Tim Adams and Don Manvel suspected that the Danaher brass were here to gauge things for themselves before asking their board to approve such a huge outlay.

Other motivations were also at work. Developing the motor had been expensive, but now Danaher owned a unique product that could sell in the millions. Dean wanted the motors to stay exclusive, especially to Ginger, so they couldn't fall into the hands of pirates.

"So even if Honda *could* take a Ginger apart and figure out how to do it," Dean was saying to the Danaher execs, "where else in the world can they get motors like yours? That's true of the motors, the amps, the gyros, the battery stacks. I think Honda will open Ginger up, and they won't call me—they'll call you. And I need to know that you'll tell them to go shit in their hat. Because you are my *sole* supplier. We're *married,* not dating, and you can't go out on a date with Honda."

Considering that Dean had been doing a little philandering of his own with Delphi, he now seemed to be adopting a monogamous view of marriage. But to prevent this particular spouse from straying, he dangled rich prospects: lucrative contracts from the Postal Service and the Pentagon. He hoped Danaher would resist the "overwhelming temptation" of a short-term seduction by Honda. Better to stay faithful while Ginger launched an industry. And oh, had he mentioned that next week he was going to see the postmaster general, who had eight hundred thousand employees and thirty thousand alternative-fuel vehicles?

H. Lawrence Culp Jr., Danaher's chief operating officer, smiled at Dean's performance and said he understood. Danaher wanted to be around for the products Dean would be making seven years from now. That sounded nice to Dean, especially since Culp was the designated successor of Danaher's current president and CEO. But Culp did seem to leave the back door open for a little fling elsewhere. And there was still the potential for being held hostage by a sole-source supplier.

"I don't care *what* pain it costs me," said Dean, "I will *not* deal with a terrorist." So did Danaher want to be his partner or just another vendor who changed prices whenever the wind blew? The executives pledged that Danaher would be a faithful partner.

But Dean didn't just want Danaher to be monogamous and honorable, he wanted an increase in financial support for another of his love children, the Stirling engine. He began the pitch with numbers: 21 million scooters were built in Taiwan the previous year; Honda sold 11 million home generators last year. His Stirling could take a big bite from these markets—and inside every Stirling could be a PacSci motor.

"We have no idea how many we'll sell or when it will be ready, but it will be a huge product in several markets," he said. If Danaher supported the Stirling the way it had supported Ginger and the iBOT, the company could be the prime supplier for all the Stirling's applications. (Note: The modifying adjective was not "sole.") And yet Pacific Scientific had just cut off its help to the Stirling project. Dean made this sound deeply unintelligent.

If doing good mattered to them, Dean added, his Stirling could make ten gallons of clean water an hour, "which would be the greatest service ever to humanity." The United Nations and the World Health Organization were excited. "And yet here is the Stirling," said Dean, "begging for motors."

At that point Dean casually mentioned that he might show the Stirling to George David, president of a little engine maker called Pratt & Whitney, who might be more generous.

"Over the next ninety days the Stirling won't be your biggest customer," said Dean, "but over the next few years—" He let Culp draw the obvious conclusion.

Culp smiled again and nodded. He promised to explore the opportunities for Danaher immediately. As the meeting began to break up, Dean joked that he liked to see Culp's company buying up so many little firms, "because we consider Danaher one of our subsidiaries."

Everyone chuckled, but from Dean's perspective it was a joke that spoke truth. The Danaher brass had come to DEKA to investigate whether Ginger was solid enough to merit a big financial commitment. Instead, Dean had put the burden on them to be dependable and honorable, and had even asked for more help. The meeting demonstrated another of his talents: the ability to deal with bigger companies from a position of weakness and prevail. He had been honest about the possible risks and benefits for Danaher, with heavy emphasis on the benefits. He had praised, cajoled, criticized, tantalized, and threatened. He had thanked his visitors for what they had already given, and then had asked for more. He was so persuasive that he usually got what he wanted. Afterward, hurrying down the hall to make his rounds, he quipped that the Danaher guys "left with a new attitude."

On the other hand, his tenacity in pursuit of the best possible deal could turn into obstinacy or fence sitting that caused delays, stress, and resentment among his suppliers and team members. He always did business his way, using nerve and instinct, and it was always personal.

Two days later, Dean threw his annual FIRST Kickoff Party, an event that grew every year. Tonight, buses moved people from local hotels to the foot of Dean's hill, where other buses took them up to the house through the wrought-iron gate flanked by tall granite plinths incised KAMEN. About thirteen hundred people were coming in two shifts. The house easily absorbed them. At the front door they were told to check their shoes and put on blue hospital booties to save Dean's polished wood floors from winter grit. It was an amusing sight, all those blue booties.

Engineers and teachers who would be working with FIRST teams made up most of the crowd, with a sprinkling of executives. They ate and drank and wandered through Dean's castle, gawking at the antique machines, the swimming pool, the helicopters, the general grandeur—and who wouldn't? They took photographs and video footage of everything, including Dean, and asked for his autograph.

At one point Dean's girlfriend, K. C. Connors, came to get me. Dean wanted me to hear the conversation about gyros between Don Manvel and J. T. Battenberg, Delphi's CEO. Battenberg asked Don how he could help. The design effort was going well, replied Don, but the price needed to come down further. Battenberg scribbled notes on a piece of paper the size of a matchbook, folding it back and forth to find empty space. Don added that BAE already had a gyro ready to go, and the Ginger team needed to make a decision next week. Battenberg said he would make some calls and send a team to Manchester in a few days. It sounded too late.

Earlier that day Tim had ordered Don to use BAE for launch. A firm, CEO-like decision. But at the party Dean told Don to give Delphi yet another chance. Don felt like he was being drawn and quartered, and he intended to avoid Tim and Dean for the next few days. He didn't think Delphi could produce what Ginger needed in time for launch, yet Dean was clearly exhilarated that this huge company was willing to try.

It was a celebratory night. Dean loved hosting this party because he loved FIRST, and he basked in the enthusiasm of these engineers and teachers devoted to it. I was feeling good, too. An hour before driving to Dean's, I had accepted an offer to do a book about Ginger. Dean pumped my hand in congratulations. So, he said, now you'll be spending more time in Manchester, right? Definitely, I said.

Four days later, everything changed.

chapter seventeen

The Leak

January–August 2001

N o one saw it coming. Even the Internet journalist who dislodged the first pebble was stunned by the avalanche that rearranged the landscape. When the dust settled, Dean and Ginger were famous. People all over the world were speculating about Ginger's identity and impatient for its arrival. Most entrepreneurs with a new product would have been thrilled by this bonanza of publicity, despite the accompanying stress, but Dean was furious and distraught, and soon banned me from the project. The following months offered glimpses of the way Dean and people around him reacted to turmoil and loss of control. Those months also changed my relationship with Dean and altered this story, briefly pushing me into the foreground and requiring a chapter I never expected to write.

On Tuesday afternoon, January 9, 2001, a reporter called from the *Hartford Courant,* my local newspaper. She wanted to verify facts that could only have come from my book proposal. My alarm bells began clanging. How did she know these things? A story about your book has just been posted on Inside.com, she said. I went to the Web site, aimed at the media industry, and began to read. Alarm turned to panic.

Someone had leaked my proposal to *Inside*'s book-biz reporter. The reporter noted Ginger's extreme secrecy but didn't hesitate to extract many of the proposal's most sensational claims and quotes: John Doerr's judgment that Ginger was the most significant technological development since the Internet, and that Dean was a blend of Edison and Ford; Steve Jobs's opinion that Ginger would be as significant as the PC and that cities would be architected around it; Jeff Bezos's description of Ginger's revolutionary nature; the investors' estimate that Ginger would make more money in its first year than any start-up in history. *Inside* wrote that I referred to Ginger as IT, a capitalized buzzword unknown to me; it turned out that my agent had used IT as a gimmick to intrigue editors. The name stuck, along with Ginger.

My first thought was Dean. He would feel exposed and vulnerable to the big companies he constantly worried about. I felt miserable for being the inadvertent cause and for the possibility that this leak might add to the Ginger team's stress. And of course I worried about how the leak might affect the book. My agent didn't know how or why the proposal had leaked, but guessed that one of the rejected editors had given it to the *Inside* reporter. I thought such things were forbidden by the ethics of publishing, as indeed they are, but that hadn't mattered. I felt naïve and stupid.

That same afternoon I mustered up the nerve to call Dean. "What the *hell* happened?" he said. It was the first time I had heard him raise his voice, but it wouldn't be the last. Who could blame him? When he understood that I hadn't leaked the proposal, he immediately calmed down, relieved. But that left everything else. We agreed not to speak to the media and hoped that the story would die quickly.

Dean said that the investors, especially John Doerr, were furious. They suspected him of leaking the proposal. Right—he had made a sudden U-turn from secretive paranoia into foolish exhibitionism. It tormented Dean that his investors could believe such a ludicrous idea, and he worried over it again and again in the coming months. The investors also blamed him for bringing a bomb—me—into the project. It was dawning on Dean that a book might go beyond what he had envisioned or could control.

Before the leak, I had intended to spend Thursday and Friday at Ginger, and I decided to stick to that plan. Best to face Dean and take my beating so we could move on. Though only two days had passed, the story had gained quick momentum, and Dean was beside himself. We spent an emotional hour in his office. Considering his anger and anguish, he was restrained and understanding. He felt caught in "a shit storm." One of his patents had been shown on national television that morning, and his inventor's paranoia had kicked in. He suspected that every engineer in America was searching his patents and that Honda already had a squad of engineers deciphering Ginger.

Steve Jobs, whose mania for secrecy matched Dean's, had called and screamed because he hadn't known that a reporter attended the meeting in San Francisco. Dean accepted partial blame for that, but since Doerr had invited Jobs and knew my purpose for being there, Dean thought it was Doerr's responsibility to inform him.

Jobs also denied the statements attributed to him in the proposal. So did Doerr, in an angry phone call, particularly his description of Dean as a combination of Edison and Ford. Doerr thought that made him sound "like a groupie." Looking chagrined, Dean said that Doerr had made the comparison many times but now was disowning it. Doerr also denied giving me permission at the San Francisco meeting to use his ecstatic e-mail about Ginger. Dean said he told Doerr, "I saw the e-mail and I saw him ask you to use it." Yet Doerr was insisting it had never happened. Like Dean, I was taken aback by this erasure of things that had been said and done and documented in my notes. This strain of amnesia turned out to be infectious.

Dean said that Doerr had threatened to pull his financing. Bluster, I argued. Doerr expected to make wads of money on Ginger, so though he might be acting stupidly at the moment, he wasn't that stupid. Dean agreed. And anyway, he added as an afterthought, he could get $100 million next week from Delphi.

The investors also had given Dean an order: Kill the book. "That's dead," said Dean. "That's over. They told me you can't have anything to do with Acros anymore."

I was stunned. Dean felt bad about it. He asked for my typical yearly income so he could prorate it and write a check for the time I'd spent on Ginger. He came up with the figure of $50,000, a fair-minded offer that I declined.

My turn to make a pitch. I said it would be a shame to abandon the book, the chronicle of an important invention. That struck a chord. Dean wondered aloud if there was a way for me to stick around, maybe covering other projects and FIRST. I said that wouldn't work. I had spent a year and a half at Ginger. That was my subject.

Dean had spent yesterday with the postmaster general. Talking about it, he smiled a little, pleasure peeking out from his clouds, and for the next few minutes became the old animated Dean, spinning out a story. During the Ginger video, when the uniformed man delivered a package, the postmaster general had said, "I get it!"

Brian Toohey and Howard Burson had been with him, and when they left the postmaster general, Burson told Dean it was the greatest sales job he'd ever seen. Dean smiled at the memory, a scrap of good news amidst the maelstrom. I said I would miss hearing stories like that. Dean looked surprised. He would miss telling them and watching me write them down. I fished his house key out of my satchel and handed it to him.

I wanted to tell the guys in Ginger that I hadn't leaked the proposal, so I walked over and key-carded myself in. They knew I hadn't leaked it, but didn't know Dean wanted to kill the book. They didn't understand Dean's logic. "Frankly, it's a mixed blessing," said Mike Ferry. Maybe now Dean would let him show the machine to consumers.

Doug Field was at home on a rare sick day. I called him to say good-bye, at least for now. He regretted Dean's verdict and didn't think the leak would damage the project at all. It might even help, he said, because it would "put a fire under some people's asses."

All this heartened me. I ran into Dean outside Ginger and asked him to keep his mind open. Killing the book wouldn't cancel the present uproar. The investors were letting their egos block their business sense. A book about the making of Ginger would add to the machine's cachet and would spread one of FIRST's primary messages—the beauty, rigor, and excitement of engineering and science. I reminded him that continuing my access to the project was his decision to make, not the investors'. He kept nodding, but couldn't figure out how to appease all the powers at play.

I remembered someone I'd forgotten in Ginger, and slid my key-card through the mechanism, as I had hundreds of times. But this time the door stayed locked. Dean had already voided me.

Tim Adams was out of the office that day, but he called that night. He thought Dean and the investors were overreacting. Dean was trying to show them that he could be tough and reestablish control. Tim didn't see much downside to letting the book continue, but he did see a downside to stopping it, because it could be valuable in the future. He intended to make this case to Dean.

In Dean's view that case was quickly overwhelmed by the pace of events. Within a week, the Ginger story had streaked around the world. The cliché "media frenzy" feels fresh when you're whirling in the midst of it. Inside.com's site crashed from the barrage of hits. My answering machine fielded calls from *Today, Good Morning America, The Early Show, ABC News,* and CNBC. Many other news organizations, including CNN and MSNBC, called the publisher or the literary agency. I stopped keeping track of the newspapers and news magazines and news services and ignored all of them. Dean had it much worse, of course. For him it was like riding the tail of a comet. He was flooded with calls and swarmed by reporters and cameramen, who loitered outside DEKA and at the gate of his estate.

My agency and publisher agreed that declining to comment was a way of showing Dean that we didn't want to make things worse for him or profit from the attention ourselves. Keeping silent wasn't easy, because the reporting was often sloppy, snide, or both. Three days after the leak, the *New York Times* ran a piece laced with errors and innuendo: The book proposal was Dean's; Dean and I were cowriting the book; the agency represented Dean, not me. Most egregiously, the reporter implied that Dean had leaked the proposal, a ploy "that would make P. T. Barnum proud."

Because the *Times* is "the paper of record," other reporters recycled and perpetuated these errors. Innuendo, repeated often enough, acquires a patina of authenticity. *Time* magazine, for example, aped the *New York Times*'s cynicism and shoddy reporting, and added its own dash of yee-haw anti-intellectualism: "Dean Kamen: Egghead gets $250K for book on secret invention. Is 'IT' a publisher-scamming device?"

Though the Inside.com reporter had stated several times that neither me, my agency, my publisher, nor Dean had leaked the proposal to him, many reporters and pundits preferred to believe otherwise. They smelled a stunt or even a hoax. Dean was no slouch at generating publicity, but such accusations infuriated and distressed him. He and I talked on the day the *New York Times* article appeared. He was considering making a statement and wanted to run it by me.

"Since speculation arising from an unfortunate, unapproved leak of a book proposal has not diminished," said his press release the next day, "I feel compelled to comment further. DEKA is currently working on several exciting projects. The book proposal referred to one. However, the leaked proposal quoted several prominent technology leaders out of context, without their doubts, risks, and maybes included. This, together with spirited speculation about the unknown, has led to expectations that are beyond whimsical. We have a promising project, but nothing of the earth-shattering nature that people are conjuring up. Please let me focus my public efforts on being an evangelist for FIRST, a cause which truly could have an earth-shattering impact."

I had decided not to give up the book. The leak, though regrettable, wouldn't keep Dean and his investors from making their billions. The epidemic of publicity had created a fever about their product that would eventually benefit them. I would keep my word to Dean by not divulging anything about Ginger until after the machine was revealed, but I wasn't going to fall on my sword for him or hand him control of my fate.

About two weeks after the leak, I tried to reach Tim Adams. His assistant acted weird. I understood why a few days later when I finally spoke to Tim: Dean had fired him. I had seen it coming, but it still surprised me, especially now, when Dean more than ever needed a steady navigator with Tim's skills and experience. But the investors had wanted Tim out for some time, and when the leak made Dean vulnerable and desperate to please them, they demanded Tim's head. I couldn't help remembering that Tim had often urged Dean to get a public relations firm in place to deal with exactly this situation, one of many pieces of good advice that Dean had ignored.

Tim didn't seem angry or bitter about getting sacked. Rather, he sounded upbeat, released. As always, his perspective was rational, not personal. He admired Dean, he said, but they just couldn't agree on how to set up and run a big new business. Tim reasoned that the investors had been looking for a way to control Dean and gain the upper hand, and the leak had provided it. Tim had advised Dean not to let them push him around—yes, they had taken a risk, but it was *his* company. Tim characterized the investors' overreaction, especially John Doerr's, as "strictly bullshit," because the public's enthrallment with Ginger allowed Doerr to reassure his partners that Kleiner Perkins's huge investment would pay off. The day after Tim's departure, Doerr had sent in a KP partner who specialized in executive recruitment to help Dean find a new CEO—who could incidentally function as KP's mole.

Tim felt frustrated that he hadn't been able to make Dean see such obvious stratagems. Tim recognized another example in the advertisement for Ginger that had started running on Amazon.com. The Internet retailer had already started taking orders for the machine, despite several rather significant unknowns: Ginger's nature, price, and date of availability. Jeff Bezos, Amazon's CEO, had the chutzpah to embellish the ad with some of the proposal's sensational quotations, including his own. He had done all this without consulting Dean. Unlike Dean, he had immediately discerned the opportunity within the leak, and as he had once advised Dean, "Business is business." (During the tumult after the leak, Bezos told a reporter, "Ginger is indeed very cool, but it is also the single most overhyped thing in the history of the universe.") Tim saw the ad as a preemptive move to capture a list of Ginger's early adopters—Dean's customers. If Dean later hesitated to let Amazon sell Ginger, Bezos could use the list as leverage.

"It's classic stuff," said Tim, "but Dean has never seen it and doesn't know what to do about it."

Tim had tried to convince Dean that the leak wasn't the end of the world, just another thing to be managed. He argued that although the publicity seemed disastrous to Dean at the moment because he hadn't been ready for it, in a year or two he would realize that in most ways it had been beneficial. Dean hated hearing that. Tim suspected that saying it had put another nail in his coffin. Dean was incapable of accepting Tim's rational analysis. Everything about the leak felt too personal.

Around the same time that he fired Tim, Dean gave Mike Ferry a choice: Work under a new marketing director or leave. Mike departed a couple of weeks later. The investors had completed a triple play—Tim, Mike, and the book were all out.

Meanwhile, the Ginger wildfire was spreading from the media into the general population. Within days of the first Inside.com story, Web sites sprang up devoted to speculation about Ginger. Most popular was theitquestion.com, which tracked more than a hundred thousand hits in its first twenty-four hours and continued at nearly the same pace for days, with participants chiming in from around the globe. Dean and Ginger broke into the Top 10 most popular subjects on the search engine Lycos, right up there with Britney Spears.

The media and the public stoked each other. Bob Metcalfe, founder of 3Com and an acquaintance of Dean's, told a national television audience, "I can't tell you exactly what IT is because I've been sworn to secrecy. . . . What I can say is, IT is a revolutionary invention that will change the world. IT's not as big as something like cold fusion, but IT is a little bigger than the Internet." Such remarks inspired more stories and more flurries on the chat sites. (Dean told me that Metcalfe hadn't even seen Ginger.) Within a couple of weeks, an Internet search turned up references to Ginger in most of the major European and Asian languages. Humor sites began running satires. Rock groups taped songs. Bible-thumpers posted scriptural allusions. It was crazy.

The frenzy reached an apogee, or nadir, in mid-February when broadcaster Art Bell promised to reveal Ginger's identity on his bizarre national radio show. His upcoming guests, he told listeners, included the publisher of *UFO Magazine,* someone doing experiments in time travel, and two people from the Ghost Investigator Society, who would share voices recorded "from beyond the grave."

Before turning things over to the guest who would reveal IT, Bell hazarded his own theory: IT stood for "inductance transportation vehicle," whose technology was a "geo-magnetic inductance neutralizing engine replacement." Then he introduced Ed Danes, "the single most controversial remote-viewer, I think, in the world." Remote viewers use psychic powers to see things far removed from them. "Neither time nor any known type of shielding can prevent a properly-trained remote viewer from gaining access to the desired target," said Danes's Web site.

Sounding confident, Danes told Bell he had remotely viewed IT. The device was indeed "a personal transportation apparatus. You hold on to it and it goes. It takes you. It's very quiet." But he must have run into some psychic static, because he also reported that Ginger moved "like a Slinky" and might sometimes levitate like a hovercraft. He wasn't sure about the hovering, because he had only allowed himself a discreet glimpse. "We didn't want to look any more, because we felt we were starting to infringe on the actual engineering," he said. "And you know, just because we have this specialized skill doesn't mean that our ethics go out the window." Dean's secret was safe with Danes.

Ginger caused such a commotion for several reasons. January is generally a sparse month for news, but the media abhor a vacuum and Ginger helped fill the news hole. (Soon after the leak, Sam Donaldson complained that the three biggest stories he had to choose from for his upcoming Sunday show were *Temptation Island,* Dennis Quaid dancing on a New York City bar, and IT.) The story also had prominent names attached, catnip to the media. And as several commentators noted, people were looking for a reason to be excited about high-tech again after the dot-com bust.

But mainly, Ginger was a mystery—a mystery catalyzed by the speed and volume of the Internet. Mystery is irresistible, because until it's resolved, anything is possible. It offers a giant screen on which to project yearnings and fantasies: crop circles, weeping statues, JFK's assassination, grainy photographs of ghosts or the Loch Ness Monster.

The mystery of Ginger was especially exciting because it came with credibility—whatever IT was, it was real. The budding hero at the mystery's center fit an American archetype, the visionary underdog entrepreneur. Dean wasn't yet well known, but his accomplishments were impressive. Famous people had spoken about him and Ginger in hyperbolic terms that invited dreams, imagination, even fantasy. The book proposal's hyped-up salesmanship—"Kamen's invention is going to sweep over the world and change lives, cities, and ways of thinking"—contributed as well. When these volatile elements combined— an empty news hole, the Internet, hyperbole, celebrity, a yearning for good technological news, a promising hero, a mystery—the story developed velocity, mass, acceleration, and momentum.

The patent drawings for Ginger appeared almost immediately after the leak, but neither they nor Dean's subdued press release kept people from indulging in wild speculation. Their fantasies turned Ginger into a device for time travel, teleportation, and magnetic levitation. Other people imagined a hovercraft, a jet-pack, an inertial thruster (IT), a mind-reading robot, and (from Thailand) "a flying super tuktuk." In the coming months, dreamers on Internet forums proposed a Bernoulli flying machine, acoustic levitation, an antigravity machine, and electrostatic levitation.

Humorists weighed in, too. Ginger was a toaster that wouldn't burn bread, an easily programmable VCR, a device that prevented Microsoft from taking over your computer, an "anti-gravity hovertoilet."

Some wag quoted the Warren Commission: "IT acted alone." Another cited Al Gore: He didn't know what IT was, but he was pretty sure *he* invented it.

Though Dean and I talked regularly, he had ordered everyone at Ginger not to speak to me. Ordinarily this wouldn't have stopped me from burrowing in, but I had grown to like many members of the team and didn't want to put them in the uncomfortable and perhaps hazardous position of disobeying Dean, especially after what had happened to Tim and Mike.

Dean and I talked by phone on February 1, just after he returned from speaking at the World Economic Forum in Davos, Switzerland. He sounded miserable. At Davos, the chairman of Sony had been asked if his company was investigating the new mystery invention, and he had said yes indeed. Dean also had heard from the FBI—they were "looking into it," or perhaps into IT, whatever "it" might be. They said they had procured "multiple copies" of the proposal from "multiple sources." To me this sounded just short of farcical, but it made Dean more gloomy.

A week later *60 Minutes II* aired a segment on him, most of which had been taped before the leak. It was a valentine. Dean's best self came across clearly: a candid, thought-provoking, eccentric engineer with world-shaking aspirations. "If you're not going to change the world," he said, "you know, go to sleep, go hibernate."

We talked the next day. The show's admiring tone had soothed him somewhat. When we spoke again on March 1, I asked if he could foresee a time when I could return to the project. No, he couldn't imagine suggesting that to Doerr or Schmertzler, at least not before the launch. I told him that the publisher wanted the book regardless.

"You can't do that," he said, agitated. "You signed a confidentiality agreement."

I assured him I wouldn't disclose anything about the machine until after it was revealed, but he had always known that I would be writing a book, which didn't violate the confidentiality agreement. The leak hadn't changed that. He grudgingly agreed.

Inside.com published another story a few days later that claimed to solve the mystery: Ginger was a hydrogen-powered scooter. Shocking news, especially to Dean and the Ginger team. Two months after the leak, the Ginger story was still hot. The *Inside* reporter spent the day hyping his hydrogen hypothesis on various talk shows, including *Today* and CNBC. That same day, the magazine *Brill's Content,* annoyed at being out-hyped by *Inside,* began hyping its exclusive interview with Dean, which claimed to dissect the media hype about Ginger.

Dean had warned me that the *Brill's* reporter seemed to be looking for a villain, and hinted that it might be me. He neglected to mention that he knew this because he had supplied him with a pitchpot full of inaccuracies. He told the reporter that the meeting in San Francisco had been mostly about FIRST, with only incidental talk about Ginger; that he believed a big portion of my book would be about FIRST; that he hadn't yet filed the patents for Ginger because the project was "still very much on the drawing board and nowhere near completion"; that he was unaware of any financial projections about Ginger; that he had never offered to pay me.

I understood Dean's motivations—to take the heat off himself and Ginger, to mollify his investors, and to put himself in the best possible light—but it was disconcerting to see him rearrange reality and erase history. Someone at Ginger had once told me that one of Dean's greatest skills was self-preservation. I didn't understand the comment at the time.

A few days later, in mid-March, Dean went on *Good Morning America* to talk about FIRST and the iBOT. He seemed to be emerging from his cave and realizing that he could benefit from the swell of publicity. One of his oft-stated goals was to turn engineers and scientists into popular heroes who could promote technological literacy. He had been handed the role, whether he wanted it or not. For instance, the Houston Forum, where he was giving a talk, was billing him as "The Nation's Top Tech Inventor, Wizard of the Mysterious 'IT' and More." He was becoming a celebrity engineer.

One afternoon in early July the phone rang. "What's cookin'?" That was Dean's customary jaunty greeting, when he was in a good mood. I hadn't heard it in months. He was chatty and friendly, though nervous about how the public might respond when Ginger was revealed, because nothing could compete with the fantasies out there. That made the launch tricky. The Ginger team had been consulting with experts who had recommended a low-key launch that would build slowly. Meanwhile, he was more and more excited by the little

Stirling engine. And FIRST—his voice always swelled on this subject—was having a growth spurt, a benefit from all the publicity about Ginger.

As always, I asked when I could return to the project. Maybe after launch, he said. Expecting the usual veto, I suggested driving up there for dinner occasionally so he could fill me in.

"We could do that," he said, "if we could be sure that no one knew about it." He didn't want us to be seen together, and suggested meeting at his house. It was Monday. He would be in Washington from Tuesday through Thursday and busy on Friday. The following week was taken up by a Ginger board meeting, but the next week his schedule got "more civilized." He said to call after that to set something up.

We spoke again on August 1. It began genially. At the board meeting, said Dean, they had decided on a low-key launch. I offered to come talk to the board. That way he wouldn't have to defend me. They could throw rocks and I could make my case. Dean said he had already thought of that. The next board meeting, to discuss media strategy, would be in late October or early November. He thought that might be a good time for "a little side meeting" about the book. But right now John Doerr and Michael Schmertzler were still bitter toward him about the leak.

I asked if he still wanted to get together. Yes, but not in Manchester, he said. Perhaps at his house in Bedford. I mentioned that a fact checker from the magazine *Men's Journal,* which was doing a long profile of him, had called my publisher and asked if Kamen and Kemper were still communicating. The publisher had said only that the book was still under contract and being written. Dean wanted them to say no comment. Too late, I said. And anyway, though the publisher had been willing to keep quiet for many months, they weren't going to deny that I was writing a book.

That's when reality hit Dean's fan.

"But you *aren't* writing a book for them," he said. "You *can't.*"

I was flabbergasted. I had been telling him for months that I was going to write a book. I told him again.

"You *never* told me you were writing a book," he said.

I felt like I had crashed through Alice's looking glass. We went back and forth a couple of times, him insisting that he didn't know, me insisting that that wasn't possible.

"Have you taken money from them?" Dean asked.

"Of course. I'm writing a book for them."

"You told me you had a contract but you weren't doing anything with it until I said it was OK." His voice had begun to rise in the harsh tone I'd gotten to know after the leak.

"No. I said I wouldn't reveal anything until Ginger was public knowledge."

"But you can't publish *anything* until we give you permission."

Another flabbergasting statement. I had never considered giving Dean such power, much less agreed to it. We had often discussed our arrangement, which gave him no control over what I wrote or even the right to review it. When introducing me to people, he sometimes joked that he didn't know what I was going to write, and might reveal his warts. Now he was inventing a new story, in which I had never told him that I was writing a book and in which he controlled anything I wrote.

His voice was calm again. He said this news would make the board very unhappy, and they might sue to stop the book. More and more bizarre. Sue to stop a favorable book about their own product? What a brilliant public relations coup.

"The agreement you signed prohibits you from writing it," continued Dean, and from revealing "*anything to do* with Ginger."

Dean was frog-kissing, inventing solutions to a perceived problem. Either that or he had believed this interpretation of the confidentiality agreement all along, and had always planned to control the book. The first explanation seemed more likely. He had been written about dozens of times and knew how journalists worked, but he was accustomed to uncritical admiration from the press. After he saw my proposal, he realized the book would include some warts, his and others'. That was unacceptable. And so now, as an expert at extemporaneous invention, he was casting around for a way to control me and everything I said about him, Ginger, and his investors. The book was a Frankenstein that mustn't escape the lab.

That was our last conversation.

The *Men's Journal* profile of Dean appeared the next month, in September. He had told the magazine's reporter that "he had no idea the proposal was being sent out." That fit the pattern of Dean's ongoing reinvention of the past and, like our final conversation, it made my heart sink. I had never known him to lie. Exaggerate to improve a story, sure, but not lie. Which meant that, incredible as it seemed, he might believe the things he had been saying.

I kept chewing on that, and eventually developed a theory. Dean sees things differently from most people. Most of us would say it's impossible for wheelchairs to climb stairs or balance on two small wheels. Most of us wouldn't dare to imagine that we could revamp world transportation or change the aspirations of American youth. But as an inventor, Dean constantly bends reality to fit his vision. If current reality doesn't suit him, he changes it. That's his habit of mind, his gift, yet it can warp into a self-protective flaw.

He once said that the laws of nature don't change, that the second law of thermodynamics doesn't depend on human approval. He loves the beautiful world of physics, the realm of fixed principles where forces follow laws and outcomes can be predicted. The leak underscored how deeply averse he is to the unexpected and the uncontrollable, and how fiercely he reacts when anything threatens his public image and his self-image.

So I can't offer details regarding the Ginger team's final strategies about media and marketing, because by then I had been purged as an agent of chaos. But I had already watched Doug Field and his engineers design Ginger, and I had watched Dean wheel and deal for its survival. That was the heart of the story, and I had it. The rest would soon be clear to everyone. Ginger was about to go public.

chapter eighteen

The Reveal

September 2001–January 2003

As the summer of 2001 turned to fall and the January 2002 launch date mentioned in the proposal drew closer, the zealots on theitquestion.com became more impatient and excited. They started predicting where and when Ginger would be revealed: over Labor Day weekend, at a high-tech conference, at the Super Bowl or the FIRST championship. Someone discovered that the raunchy television cartoon *South Park* was running an episode in November 2001 called "The Ginger Device" (later renamed "The Entity"). Maybe Dean was using the show to offer a clue? Evidently not: The episode featured a gyroscopic vehicle powered by several penis-like pistons inserted into the rider's orifices.

After the terror of September 11, 2001, someone on theitquestion.com started a thread entitled "Help Us Dean Kamen." When a Defense Department official mentioned that the United States was deploying secret weapons in Afghanistan, a poster on the site speculated that this referred to Ginger. Someone else suggested that the special operations troops in

Afghanistan were hunting Osama bin Laden on Gingers. Another poster correlated the drop in OPEC's oil prices after September 11 to the invention's imminent release. And had anyone noticed that British prime minister Tony Blair, while speaking about terrorism, had declared that science could solve the world's problems? Obviously a veiled reference to Ginger, noted another poster. When George W. Bush predicted an economic turnaround by year's end, a chat-site fanatic deduced that the president had inside information that Ginger would arrive soon.

In short, all Dean and Ginger had to do was save the country, win the war, defeat international terrorism, and, incidentally, rescue the economy. Preferably through levitation or teleportation.

Then, on November 26, Diane Sawyer announced on *Good Morning America* that IT would be revealed on the show in one week, and the speculation and hype cranked up again. On the morning of the reveal, when the curtain lifted off the machine and Diane Sawyer said, "But that *can't* be it," she spoke for many. The show's hokey, breathless presentation turned one of Ginger's greatest virtues—its simplicity of design—into a negative. Standing still, the machine looked ordinary, even mundane.

The mood changed as soon as Sawyer and cohost Charles Gibson took Gingers to nearby Bryant Park for a ride. Gibson's face turned eloquent with delight. "This is *really* cool," he said, yukking like a teenager. Sawyer, usually an ice goddess, melted to the point of doing tricks, riding with no hands and lifting one leg backwards. She even let Dean run over her foot. "I don't feel a *thing*," she gushed. "It's *astonishing*."

As Dean and the hosts tooled around the park, I glimpsed many members of the Ginger team: Doug Field, Mike Martin, Ron Reich, Bill Arling. Industrial designer Scott Waters and controls engineer John Morrell had been assigned to shadow Gibson and Sawyer to prevent them from injuring themselves during Ginger's debut. Scott smiled and looked relaxed, but John, as always, looked worried. All of them except Dean wore homogenous outfits of khaki pants, light blue dress shirts, and dark blue vests. It was odd to see their idiosyncrasies forced into a uniform chosen by the marketing department.

Near the end of the show, the group returned to the studio. J. D. Heinzmann stood in the crowd. The hosts invited another of that day's guests, the rap producer Russell Simmons, to step onto a Ginger. "Oh my God!" said Simmons, his face lighting up. "This is so cool!" Rolling back and forth, he asked Dean twice if he could invest or buy stock. "Help us get our FIRST program into schools," said Dean.

The machine looked just like the last version I had seen. There were a few surprises. Though the logo hadn't changed, the name Flywheel had been replaced by Segway, a clumsy solecism. (In the coming months, many commentators preferred to call the machine by its code name, Ginger). Dean had told me in August about the slow commercial launch, but I hadn't expected the consumer launch to be postponed for almost a year, to late 2002. (It was later extended further, to sometime in 2003.) The price was startling, too. According to published reports, the commercial model had jumped to $8,000 to $10,000, the consumer model to about $3,000.

(These early projections about availability and price turned out to be overoptimistic. Just a year later, in mid-November 2002, Dean and Jeff Bezos would appear together on *Good Morning America* to announce that a consumer model finally could be ordered on Amazon—for $4,950, with shipping sometime between March and July 2003. As Bezos had suggested two years earlier in the DEKA cafeteria, Segway and Amazon required buyers to send a 10 percent down payment to secure their places in line, thus providing Dean and Bezos with an infusion of interest-free capital.)

Immediately after the reveal, Dean wrote a gracious letter to the members of theitquestion.com, commending them for their "dedication, stamina, and Internet research skills." He admitted that the real Ginger couldn't compete with the forum members' imaginations, but he compared the machine to other world-transforming inventions that hadn't been accepted overnight—the automobile, alternating current, and the airplane.

The letter raised Dean in the estimation of some forum members, but after eleven months of feverish and often outlandish speculation, others snickered at Ginger, mocking it as an expensive scooter, "a two-wheeled Dorkmobile." Some of the thread titles tell the story: IT is Ridiculous. A Catalyst for Change. Stupid, Overrated, and Way Over-Hyped. I'd Rather Walk. You Guys Lack Vision. That Wasn't IT.

The reveal had been expertly orchestrated. Burson-Marsteller, the PR agency, arranged a sneak preview of Ginger for the *New York Times,* which wrote a long, complimentary article. The agency also packaged

Dean, Ginger, and a handpicked reporter with a connection to John Doerr as an exclusive for *Time* magazine. (The same reporter later wrote in *Vanity Fair* that Dean required him to sign "a prodigious confidentiality agreement.") *Time* went for the story in a big way, with a seven-page spread entitled "Reinventing the Wheel." Dean told the magazine that Ginger "will be to the car what the car was to the horse and buggy," and John Doerr predicted that the new company would reach $1 billion in sales faster than any venture in history. No more subdued disclaimers.

Stories about Dean and Ginger flooded the media, as much because the mystery had been solved as because of the invention itself. *Good Morning America*'s Web site registered more than a million hits on its segment. Burson-Marsteller reported that all the major networks and many foreign networks covered the story in the first few days, as did scores of print media. More than fifty national and international media outlets sent photographers to a photo call. Burson-Marsteller, stretching credulity, claimed an audience of more than 113 million people for its video news release. Segway reported thousands of hits per minute at its Web site (according to the company, 4.7 million people visited the site in less than seven weeks; nine months later, Dean claimed 59 million hits on the site). AdAge.com reported that Segway generated 758 million "impressions" in December, with an advertising value of $70 million to $80 million, followed by similar numbers in January and February. On the search engine Lycos, Segway jumped to 3 on the Top 10 list, behind Christmas and Dragonball but ahead of Britney Spears and Harry Potter.

Newton's Third Law states that for every action there is an equal and opposite reaction. That proved true of public opinion as well, which ranged from IT's Fabulous to IT's SHT. (The pricey naming experts evidently missed the problematic acronym formed by Segway Human Transporter.) Many people were disappointed, even bitter, that Ginger's appearance smacked more of the Flintstones than the Jetsons. "It won't beam you to Mars or turn lead into gold," Dean told *Time*. "So sue me." Some reporters compared it to a push-mower. "Those credulous investors who anointed inventor Dean Kamen the next savior of the technology sector," said *Forbes*, "must be feeling a little foolish today." Some commentators derided Ginger as a toy for rich people, a criticism that had been leveled at Edison's phonograph and the first automobiles. To Dean's chagrin, nearly everyone called Ginger a scooter, a word that didn't go well with "revolutionary" or "sidewalk."

Ginger's image began showing up everywhere. Editorial cartoonists had a heyday, drawing the machine with seats, fins, and combustion engines, or placing it as the last step in human evolution. The *New Yorker* ran a cover of Osama bin Laden and his cronies escaping Afghanistan on Gingers. On eBay, a painter who described himself as an "outsider hillbilly artist" offered an acrylic entitled "Jesus On A Segway."

Soon after the reveal, Jay Leno rolled out to do his monologue on a Ginger. Dean was a guest that night, sandwiched between Russell Crowe and Sting. Leno, Crowe, and Sting all wanted to ride Gingers, so Dean scuttled alongside with his arms spread, fretting that they might crash or drive off the stage. Two nights later, David Letterman chattered about the famous invention, then asked its inventor to come on stage. The "inventor," a goofball in a billed cap, said, "It's a piece of crap with a whirligig that'll set you back about five G's. Come on, losers, make me a billionaire."

Some commentators were nastier. In his predictions for 2002, techno-pundit John Dvorak, who had accused Dean of shameless self-promotion after the leak, wrote that the machine would never make money, "So the only way the investors can get out alive is to sucker the public into buying stock." The magazine *Business 2.0* included Segway as one of its "101 Dumbest Business Moments," comparing it unfavorably to a $100 eighteen-speed bike. The McLaughlin Group, that academy of high-tech wisdom, gave Dean its award for Most Stagnant Thinker.

Yet when MSNBC asked people to vote for the top technology of the year, Ginger came in second behind wireless phones, and CNN called Ginger the top tech story of 2001.

Dean's fame grew with the coverage. By mid-April he merited attention in that barometer of celebrity, the newspaper gossip page. The *New York Post* reported in its "Sightings" column that Dean was spotted "scooting round the cozy confines of restaurant Two Two Two on West 79th Street on his self-balancing, electric-powered machine."

Dean campaigned hard, traveling the country to lobby officials and reporters. Skeptics often turned into believers after a blast of Dean's personality and a spin on the machine. In the lion's den of Detroit, the reporters couldn't wait to chomp on him. One of them accused him of creating something that would make Americans even more sedentary. Dean responded that the Segway was a labor-saving device that would open up more time for walking and exercise, then asked if the reporter was avoiding a great opportunity for exercise by using a washing machine instead of pounding dirty laundry with rocks.

When one reporter accused Ginger of being dangerous to pedestrians, Dean drove straight into the guy, who was surprised but unhurt. Another reporter proclaimed that the machine would never catch on in Michigan because of the icy winters. Dean replied that while testing the machine on snow and ice in New Hampshire, the engineers on Gingers would grab the ones who were sliding around and pull them like this—whereupon he grabbed the reporter's wrist and backed up, towing her with him.

The next day, a columnist for the *Detroit Free Press* grumbled, "I'm paid to be skeptical. Kamen's unanimously flattering publicity and the relentless media oohing and ahhing over the HT struck me as too contrived, too good to be true." Then he took a ride. "My skepticism evaporated the moment I stood on the HT," he wrote. "This thing is so cool that all I can say is the exuberant hype it's received so far is understated."

Similar stories appeared wherever writers stepped onto Ginger. A reporter for smartbusiness.com wrote, "All it takes to be convinced is one ride. Try it and you'll see why so many billionaire CEOs are giddy about this thing. It's the perfect middle ground between feet and cars."

When President George Bush visited the University of New Hampshire, Dean engineered another publicity coup. He showed up at the reception on a Ginger and persuaded Bush to ride it in private—and then told the press all about the president's delight.

Officials and legislators who rode Ginger surrendered the same way. Brian Toohey, charged with getting the machine onto sidewalks, had done his job well. Earlier in the year, he and his Washington contacts had convinced the National Highway Safety Administration that Ginger was not a vehicle and therefore shouldn't be regulated by the NHSA. Similarly, the Consumer Product Safety Commission agreed to classify Ginger as a consumer product. The public might perceive Ginger as a scooter, but these federal regulatory agencies had agreed not to. The rulings allowed the company's lobbyists, hired in dozens of states before the reveal, to argue that the machine should be exempted from local regulations that might keep it off sidewalks.

Toohey sent teams to dozens of state capitals to demonstrate Ginger. Sometimes Dean showed up to emcee and sell the legislators. By fall of 2002, thirty-two states had passed special laws or exemptions that permitted Ginger on sidewalks, including a crucial win in California. It was surprisingly easy. The company hoped to make Ginger legal in forty-four states by the end of 2002 and in the other six states when their legislatures met in 2003.

The regulatory campaign's second strategy aimed to put public employees on Gingers. Both strategies had the same goal: to get the machine accepted on sidewalks. If cops and mail carriers rode Gingers there, how could the general public be prohibited when the product was available to everyone?

The Boston Police Department tried out some Gingers on New Year's Eve. According to the *Boston Globe,* the rolling cops drew "a deafening cheer" from the crowd celebrating First Night. The police departments in Chicago and several other cities also experimented with Gingers. The U.S. Postal Service tested the machine on routes in places such as Tampa; Memphis; Concord, New Hampshire; and San Francisco, where it easily handled the hills. The testing went well, and the Postal Service bought forty machines for a second phase of tests. The National Park Service tried out Ginger in Washington, D.C., and along the south rim of the Grand Canyon.

The most enthusiastic early adopter was the city of Atlanta, whose officials called Dean after learning about him through the leak. The city offered itself for a pilot program and put Gingers into use by its police department, downtown Ambassador Force, and airport personnel.

Not everyone surrendered to Ginger. Aside from questions about discomfort in cold or wet weather, most objections fell into two categories: health and safety. "If [Ginger] becomes popular for otherwise healthy people," said Dr. Philip Ades, director of preventive cardiology and cardiac rehabilitation at the University of Vermont College of Medicine, "it will rank with electric garage door openers and automatic car door openers as an absurd extension of laziness and slothfulness that will further increase levels of obesity and heart disease in America."

That sort of overstatement could be easily dismissed, but the safety issue was more nettlesome. A number of groups expressed concerns about collisions between pedestrians and machines, and also voiced other nebulous yet valid worries. For instance, people might feel less comfortable on sidewalks if Segways were buzzing by them. Sidewalks were already too crowded—and who had a greater right to be there, walkers or riders? Would letting the Segway onto sidewalks open the door for more dangerous motorized devices such as scooters?

Several consumer groups urged people to protest to legislators about legalizing the machine on sidewalks. The American Academy of Pediatrics asked Congress to require an age limit, licensing, and equipment such as helmets. Though no one had been seriously injured by a Segway (one rider in Atlanta fell off and hurt a knee), senior citizens in San Francisco protested against the machine with signs that said, "Stop the Segway Slaughter" and "Segway: Zero Emissions, Senior Killer."

Some of this can be ascribed to the knee-jerk instinct of worry-warts and bureaucrats when faced with something new or unknown: Tie it up, rule it out, fence it off, shut it down. New technology, in particular, has always been regarded with suspicion, from Galileo's telescope to television. Someone on theitquestion.com pointed out that the first car owners were required to stop at every intersection and wave a flag or honk the horn.

But Dean's assurances aside, what *would* happen if an eleven-year-old on a skateboard collided head-on with a 200-pound man on a Ginger? If a rider going 12 miles per hour rear-ended a senior citizen, would the oldster agree that it was no worse than a bump? No one knew, because no one had Gingers, which left room for alarm. After the initial quick legalization by a couple of dozen states, a mild backlash began. A few state legislators expressed doubts about letting the machine onto sidewalks, and a few local officials demanded the right to set their own regulations. San Francisco banned the machine on its sidewalks. But considering how new and potentially disruptive Ginger could be, Dean and Brian Toohey must have been delighted with how quickly the machine advanced through the legislative and regulatory brambles.

The machine's projected consumer price of $3,000 to $4,000 raised another set of criticisms. A few commentators pointed out that people often paid more than that for computers, snowmobiles, and other expensive toys, but the majority thought that $3,000 would make the machine a rich person's toy. As Jeff Bezos had told Dean a year earlier, the price of the consumer model would have to drop like a stone if Dean wanted to sell enough machines to revolutionize transportation. Yet by November 2002, when Dean and Bezos jointly announced the consumer model, Bezos had changed his tune, telling millions of people on *Good Morning America* and CNN that $4,950 was a rock-bottom price that wouldn't be going down.

Nor was it clear that large commercial businesses were going to embrace Ginger. John Doerr was predicting that the new company would reach $1 billion in sales faster than any start-up in history, but sales took off like a tortoise. In the months after the reveal, only a few companies, most of them suppliers for Ginger, tested the machines at plants and warehouses: Michelin, GE Plastics, Delphi, Amazon. Businesses tiptoe toward large purchase orders, so it's too soon to predict whether Ginger will succeed in that arena, but six months after the commercial launch, the new company was nowhere near Bob Tuttle's estimate of fifty thousand to a hundred thousand sales in the first year.

During May, sweeps month on television, Ginger wheeled its way into popular culture. Niles, the snooty character on *Frasier,* got a machine and loved it so much he wouldn't let anyone else ride. Andy Rooney, the *60 Minutes* pundit, rode Ginger at an auto show and said he wanted one for Christmas. Delphi began running television ads in which a Ginger streaked by in the background. That same month, the actor Peter O'Toole cut an aristocratic figure riding a Ginger on the show *The Education of Max Bickford.* He played a professor attempting to lure Max Bickford (Richard Dreyfus) to Harvard.

"You think forward and off she goes," said O'Toole's character, rolling. "You think stop and she does. Revolutionary."

"I've heard of it," said Bickford. "How did *you* get one?"

"A colleague of mine at Harvard invented it. Dean Kamen. Cutting edge."

Well, most of that was accurate.

Ginger figured in an episode of *The Simpsons* during the same week that *Meet the Press* showed a news clip of Vice President Dick Cheney riding the machine. Cheney said his was a loaner from Dean, who had also invented the stent in his heart. For Ginger, May was quite a month.

That same month's issue of *Vanity Fair,* purveyor of what's hot, chic, and scandalous, devoted more words to Dean than to its other lead subjects, John F. Kennedy Jr. and the newest starlet. The principal investors' comments about the leak were entertaining, considering how they had reacted at the time. Michael Schmertzler dismissed the damage as "minimal." John Doerr, who with Steve Jobs had howled loudest at Dean, now effused that the leak "actually had an enormously salutary effect on the business." Though the leak had given Dean, Ginger, and FIRST immense amounts of attention and was the reason that the debut of a machine used by a few police officers and postal workers had received worldwide hype, Dean still insisted on unmitigated calamity. "I can't think of anything good about that event," he said. "It was the single worst thing that has ever happened to me in business."

The story also revealed that John Doerr and Michael Schmertzler still wanted to invest in the Stirling project, now called Power Concepts, but hadn't yet offered terms Dean would accept. Dean was still running Ginger himself. John Doerr's handpicked replacement for Tim Adams hadn't survived a two-month audition, and Dean had vetoed all of Doerr's subsequent suggestions. Sixteen months after Tim's departure, in the same month that the *Vanity Fair* story appeared, Dean finally hired George T. Muller, former president of Subaru of America, as president of Segway. Dean kept the titles of chairman and CEO for himself.

According to the article, he continued to frustrate his investors in familiar ways. Federal Express had pulled out of a large deal, exasperated because Dean wouldn't commit to a firm price. Late into 2001, Segway still had done little market research and no field testing, nor had Dean hired a director of sales or any salespeople. Clearly, Mike Ferry had been blamed for a lot of deficiencies that didn't originate with him. Tobe Cohen had been named marketing director.

The article didn't mention that a few months before the reveal, Don Manvel, Ginger's director of manufacturing operations, had resigned, worn out by Dean's dodgy tactics with suppliers. Remember the gyro ruckus, when Dean whipped Don back and forth between BAE Systems and Delphi in hopes of saving money? Ginger had entered the world balancing with BAE gyros. Evidently Don, Tim Adams, and Doug Field had been right that Delphi couldn't produce a timely House of Gyros.

The article noted that many months after Mike Ferry's departure,

Dean had finally filled the position with Gary Bridge, a top marketing executive from IBM. The article described Bridge emerging from "an extended session" with Dean, looking as if he had "been beaten with a stick" and complaining that such sessions kept him from getting much done. By the time Dean and Jeff Bezos announced the consumer model in November 2002, Bridge was gone, another casualty, as was a short-lived director of commercial sales. Tobe Cohen, Segway's main marketing spokesman since the reveal, also had disappeared. The high rate of executive turnover continued in January 2003, when George Muller resigned after just a year and a half as Segway's president. Vern Loucks, Dean's old friend, who had invested in Ginger and also sat on the board, took over as CEO.

I wondered about the rest of the team. It was easy to imagine J. D. Heinzmann happily absorbed in some new electronic tangle or proselytizing for his latest avant-garde design. John Morrell would be in fruitful turmoil over Dean's push for new, less expensive versions of Ginger, and fretful about taunting the gods as he worked on the math to make them safe. Scott Waters would be agonizing over the looks of Ginger's next generation, beautifying it a few square centimeters at a time. Ron Reich would be at his elbow, barking about manufacturability. Doug no doubt was still working twice the normal hours and pondering ways to infuse his growing team with the mission of preserving the soul of this new machine.

It wouldn't be easy. To help me understand what he was trying to accomplish with his engineers, Doug once recommended a book by Warren Bennis and Patricia Ward Biederman called *Organizing Genius: The Secrets of Creative Collaboration*. The book analyzed the dynamics of what the authors called "Great Groups" at Disney Animation, the Manhattan Project, Xerox's Palo Alto Research Center, and Lockheed's Skunk Works. The Ginger team shared many characteristics with such groups. It was nonconformist and, except for occasional Deanings, nonhierarchical. Its members were driven by the certainty that Ginger was more a calling than a job. The group had a charismatic, visionary leader in Dean and a pragmatic, organized dreamer in Doug, both of whom instilled the team with the conviction that they could change the world.

But *Organizing Genius* also warned that Great Groups often collapsed when the project ended and exhilaration dwindled into routine.

That's why Doug had worried that once Ginger was finished, Dean might relegate his engineering team to servicing the assembly plant instead of devising newer, more exciting versions of the machine.

Whatever the future held for them, I was sure that Dean, Doug, and the engineers felt proud of their technological feat. They had started with an unsightly, hazardous prototype, and in less than two years had crafted a sleek, exciting, safe machine that could be mass-produced. People who scoffed at Ginger as plain or dorky were ignorant about engineering, design, or both, and obviously had never ridden it. Scott Waters's qualms must have been eased a bit in July 2002 when the machine received a gold medal in the Industrial Design Excellence Awards, cosponsored by the Industrial Designers Society of America and *Business Week*. In April, Dean won the Lemelson-MIT Prize for Ginger, a splendid honor because it's the only prize specifically for inventors. Dean pledged the $500,000 award to FIRST. Beyond all the hype and politics surrounding Ginger, it seemed clear that people who understood the machine were astonished by it.

Ginger also took Dean closer to another of his goals—turning engineers and scientists into cultural heroes. In late May, under the headline, "Great Minds, Great Ideas," *Newsweek* magazine asked a question that would have been inconceivable two years earlier: "Will scientist Stephen Wolfram and inventor Dean Kamen be known as the twenty-first century's most important thinkers? If all goes according to their plans . . ." The article called Dean "the poster boy for invention" and described his world-shaking ambitions.

It also noted that he came home to an empty house. (He had told his longtime girlfriend, K. C. Connors, to move out before the reveal.) "I can start the biggest [technical] project in the world," Dean told *Newsweek,* "but I think getting in relationships is riskier and to me scarier."

In what was becoming a common trope, the article compared him to Thomas Edison, epitome of the lone American inventor plotting in his lab to change the world.

Castles in the Air

"To believe your own thought, to believe that what is true for you in your private heart is true for all men—that is genius," wrote Ralph Waldo Emerson in his essay "Self-Reliance." Dean Kamen fits that mold. His willingness to pursue his ideas with his own money has brought him wealth, awards, and fame. His accomplishments stem as much from his stubborn self-reliance and his salesman's confidence as from his engineering ideas. He often told me that he needed to retain control of his inventions because investors and stockholders want to turn a profit, not change the world. They wouldn't let him spend millions of dollars exploring wild dreams that might lead to an iBOT, a Ginger, or a Stirling engine. Investors can't see the prince in the frog.

That's why the world needs inventors. Major innovations change the way people see the world. The railroad altered perceptions of space and time, which were altered further by the automobile and the airplane. The telegraph revolutionized communication, the telephone did it again, and the cell phone and Internet began a third wave. The electric light, the atomic bomb, the personal computer—all changed the way we think as well as the way we live.

Does Ginger have that power, as Dean believes? It's too soon to tell. The marketplace is littered with the bones of worthy engineering ideas. Ginger is, in the current catchphrase, a "disruptive technology," so it faces more than the usual number of booby traps. The new company might have to fight powerful opponents with deep pockets. It almost certainly will get wounded in expensive battles over patent infringements by Asian pirates or multinational corporations. It might have to fend off claims by other patent holders. A month after the reveal, a Japanese engineer asserted a prior patent for Ginger's basic technology. This machine has never been built, but unless the claim is resolved it could keep Ginger out of the huge Japanese market. (Even if Dean's worst nightmares about pirates and patent-jumpers come true, he counted on one consolation. "By the time Edison died, there were lots of people making light bulbs," he once told me, "but no one remembers *their* names.")

If Ginger is to alter cities, it will need lots of help, beginning with an infrastructure of rental kiosks, charging stations, and service centers. It will require architectural modifications to subway stations and other high-density areas. Other changes will be more subtle. A new definition of pedestrian will have to evolve, as well as a new etiquette on sidewalks.

The biggest wild card is the public. "In order for invention to fulfill its social function it must be recognized," wrote Norbert Wiener in *Invention: The Care and Feeding of Ideas*. No one knows how consumers will respond to the machine. They might shrug or marginalize it as a toy. They might refuse to pay the asking price. They might not recognize how the machine can fit into their lives. They might decide that the niche Dean intends for Ginger, between walking and driving, doesn't need filling.

Even success poses dangers. If hundreds of thousands of consumers want a Ginger in the first year, as the company projects, can the factory meet demand? If not, how resentful will people be? If manufacturing does keep up with demand, how will people respond when hundreds or thousands of Gingers begin rolling down their city sidewalks? Accidents are inevitable, as are lawsuits. Too many of either could be financially disastrous for the company, and might also push legislators and regulators to change their minds and ban the machine from sidewalks.

Dean sweeps all this aside with his usual confidence. "I would stake my reputation, my money, and my time on the fact that ten years from now, this will be the way many people in many places get around," he told *Time* magazine.

If the marketplace decides that Ginger is a minor innovation rather than a revolution, the company will likely still make billions on commercial applications, but Dean will consider that a failure.

"The reason that we uncharacteristically decided to build a whole company around [Ginger]," he said at the Harvard Business School's Cyberposium in February 2002, "wasn't because in the end we would make a lot more money, although of course we will if it succeeds. It's because I couldn't see spending years of my life developing something that would turn out to be a scooter." If you want to change the world, you can't leave it up to somebody else.

But Dean himself is one of Ginger's potential booby traps. He doesn't know how to run a large product-development and manufacturing company, yet is reluctant to give power to anyone else. He constantly undercut Tim Adams and quickly dismissed Tim's replacement. Even if the company's current president is exceptional, history and habit suggest that Dean won't hand him the reins. Tim Adams used to hope that Dean would relax his grip once the machine went into production, but relaxation isn't part of Dean's character.

"If I'm awake, I'm working," he often says. For years he has been running himself ragged to develop and finance the iBOT, Ginger, the Stirling engine, and FIRST, not to mention DEKA's other projects. Now Ginger is poised to become an international company, and if his ambitions for the Stirling engine ever ripen, his newest spin-off, Power Concepts, will follow the same pattern. The much-delayed iBOT is inching down the pipeline, too, and the engineers at DEKA are never idle. Dean is no longer directing one small R&D company whose books can be handled by his mother. If he keeps trying to micromanage everything, he might, in engineering parlance, run out of headroom.

His new celebrity also will be a test. He often declares that technologists deserve to be as well known as athletes and actors. Now he's on the verge of that himself. He wants fame the way he wants everything else—

on his terms. But celebrity isn't a negotiation he can control. Once you put yourself into the limelight, you will be lit from all angles. Perhaps even by a book. Dean will have to make peace with that.

In *American Genesis,* a history of invention and technology, Thomas P. Hughes writes that after an innovation left the lab for the market-place, "the inventor could no longer order, systematize, and control his or her brainchild." Other forces take over: public opinion, regulation, competition. Like any inventor, Dean is fanatical about filing patents for as many applications of his technology as possible. But other entrepreneurs, some of them pirates and patent-jumpers, will find unforeseen uses for Ginger and its technology. Dean won't be able to control it. As Steve Jobs warned him in San Francisco, he will wear himself out if he tries to. But he might try.

As Henry David Thoreau advised in *Walden,* Dean builds castles in the air and then puts foundations under them. Like earlier brilliant inventor-entrepreneurs such as Robert Fulton, Samuel Morse, James Eads, Andrew Carnegie, and Thomas Edison, Dean has a flair for sales-manship and promotion equal to his gift for hands-on invention and engineering. That combination can shape the future. But in recent years Dean's focus on promotion and management has pulled him away from his gifts as an inventor. He often complains that he misses engineering, but to return to it would require giving up some control over his expanding businesses. Inventors, not managers, change the world.

At the beginning of the Ginger project, even the machine's most enthusiastic supporters couldn't envision what Dean imagined. One day in the lab, Doug Field reflected on that vision. "Dean saw *all this,*" Doug said, patting his team's newest Ginger, "in *that.*" He pointed at Mary Ann, the rough prototype that first captured his engineering heart. "Amazing," said Doug, shaking his head. "I saw a recreational vehicle. A little company. It's good that I wasn't in charge, because I would have taken it small-time. But Dean *saw* it."

Dean told the audience at the Harvard Cyberposium that he hoped Ginger, the iBOT, and the Stirling engine each would find the place he had imagined for it, because "it will make the world a better place. It will certainly make us richer. But the richer part will be a byproduct of the fact that we did something important. And in each case, it will probably have taken five or ten years. Which I happen to think is a short time for a really big idea to happen."

Big ideas and the future: an inventor's stock-in-trade. Dean once said that science is about why, engineering is about why not. Why not a revolution in urban transportation? Why not a revolution in energy delivery? Why not clean water for the world's masses? Why not a transformation of America's youth and culture? "Give me a lever long enough and a fulcrum on which to place it," said Archimedes, one of Dean's heroes, "and I shall move the world."

People sometimes ask what Dean considers his greatest invention. He usually says FIRST, because it inspires teenagers whose innovations might someday benefit humanity. But his FIRST love is hardly his last. Dean intends to keep engineering the future.

"I don't consider this the end," he told a newspaper after Ginger was revealed. "This is the end of the beginning."

With minor exceptions, such as personnel reviews and legal discussions, I had full access to Dean Kamen, the Ginger team, and other projects at DEKA from August 1999 to January 2001. From January to August 2001, I spoke to Dean periodically by phone.

My primary source has been more than 5,000 pages of notes, most of them made during firsthand observation of the events described. Dean and I often discussed the rules of journalism that governed my relationship to him and the project, and we explained them to the Ginger and iBOT teams at a preliminary meeting. The essential rule: Everything is on the record unless specifically requested otherwise. Such requests were exceedingly rare. I took notes openly, whether in the lab, in meetings, or at dinner with Dean. I have not altered chronology, invented scenes or details, or fabricated dialogue and quotations. In the text, extended dialogue without quotation marks indicates that the conversation was relayed to me by one or more of the participants rather than witnessed.

Secondary sources include several hundred pages of documents given to me during meetings of the project's managers, engineers, marketers, investors, and board of directors, as well as a long shelf of books about engineering, invention, and entrepreneurship. I have tried to tell the truth; any inaccuracies are unintentional.

My publisher provided the manuscript of this book to Dean Kamen to allow him the opportunity to identify for our consideration confidential information about DEKA's finances, products in development, or other matters that he regarded as confidential or sensitive. Some of the information so identified has been redacted.

ACKNOWLEDGMENTS

Dean Kamen, for the nerve and imagination to invite me into his business and his mind.

The members of the Ginger team and others at DEKA, for their acceptance and patient explanations.

Hollis Heimbouch, for championing this book at Harvard Business School Press, and Constance Hale, for improving it through spacious comprehension, acute criticisms, and astute suggestions.

Rafe Sagalyn and Dan Kois, for seeing the possibilities and riding the storm.

Jim Doherty, for sending me in 1994 to check out a character named Kamen, and for a decade of cantankerous editorial support.

Nancy Aronie, for pleasingly premature enthusiasm.

Ben and Alex, for daily restoratives of joy.

Jude, for unflagging faith and encouragement, and (almost) bottomless tolerance.

Steve Kemper is a journalist whose work has appeared in *Smithsonian* and the *National Geographic* magazines, among many others. He holds a Ph.D. in English from the University of Connecticut. Kemper lives in West Hartford, Connecticut, with his wife and two sons. *Code Name Ginger* is his first book.